CELL PÓPULATIONS

Vol. 8
in the Series now titled
METHODOLOGICAL SURVEYS – *sub-series* (B): Biochemistry

Series Editor: Eric Reid
Wolfson Bioanalytical Centre

University of Surrey

Other titles, with retrospective assignment to the two sub-series

Below: Sub-series (B): Biochemistry

1971 SEPARATIONS WITH ZONAL ROTORS
[ranks as Vol. 1]
published by (and available from)
the Centre at Guildford

METHODOLOGICAL DEVELOPMENTS IN BIOCHEMISTRY
(Series title, now superseded by METHODOLOGICAL SURVEYS)
published by Longman Group Limited (Vols. 2-4; available only from the Centre)
or North-Holland Publishing Company (Vol. 5)

1973 SEPARATION TECHNIQUES (Vol. 2)
1973 ADVANCES WITH ZONAL ROTORS
(Vol. 3)
1974 SUBCELLULAR STUDIES (Vol. 4)

Sub-series (A): Trace-Organic Analysis
1976 ASSAY OF DRUGS AND OTHER
TRACE COMPOUNDS IN
BIOLOGICAL FLUIDS (Vol. 5)

METHODOLOGICAL SURVEYS [IN BIOCHEMISTRY – *optional phrase, now superseded]*

1977 MEMBRANOUS ELEMENTS
AND MOVEMENT OF MOLECULES
(Vol. 6; including muscle)
1979 CELL POPULATIONS
(Vol. 8: *this volume*)
1979 PLANT ORGANELLES (Vol. 9)
1981 (intended) CANCER-CELL
ORGANELLES *(Vol. 11)*

1978 BLOOD DRUGS AND OTHER
ANALYTICAL CHALLENGES
(Vol. 7)
1980 (intended) SAMPLE HANDLING IN
TRACE-ORGANIC ANALYSIS
(Vol. 10; including environmental)

CELL
POPULATIONS

Edited by
ERIC REID, PhD, DSc
Director, Wolfson Bioanalytical Centre
University of Surrey

ELLIS HORWOOD LIMITED
Publishers   Chichester

Halsted Press: a division of
JOHN WILEY & SONS
New York - Chichester - Brisbane - Toronto

First published in 1979 by

ELLIS HORWOOD LIMITED

Market Cross House, Cooper Street, Chichester, West Sussex, PO19 1EB, England

The publisher's colophon is reproduced from James Gillison's drawing of the ancient Market Cross, Chichester

© 1979 Eric Reid/Ellis Horwood Limited

British Library Cataloguing in Publication Data
Cell populations. – (Methodological surveys in biochemistry; vol. 8)
    1. Cell populations.
    I. Reid, Eric II. Series
    574.8'766   QH587   79–40729
ISBN 0–85312–120–6 (Ellis Horwood Ltd., Publishers)
ISBN 0–470–26809–3 (Halsted Press)

Printed in Great Britain by Unwin Brothers Limited of Woking

Distributors:

Australia, New Zealand, South-east Asia:
Jacaranda-Wiley Ltd., Jacaranda Press,
JOHN WILEY & SONS INC.,
G.P.O. Box 859, Brisbane, Queensland 40001, Australia.

Canada:
JOHN WILEY & SONS CANADA LIMITED
22 Worcester Road, Rexdale, Ontario, Canada.

Europe, Africa:
JOHN WILEY & SONS LIMITED
Baffins Lane, Chichester, West Sussex, England.

North and South America and the rest of the world:
HALSTED PRESS, a division of
JOHN WILEY & SONS
605 Third Avenue, New York, N.Y. 10016, U.S.A.

Editor's Preface

with comments on nomenclature

Cell isolation may seem an odd theme for a 'Subcellular Methodology Forum' (Guildford; July 1978) whilst clearly worthy of the desk-book form into which the material has now been hewn. Those concerned with isolating subcellular fractions are, however, increasingly aware of the problem of cell-type diversity that may imply biochemical diversity, and of the desirability of separating from their starting material a particular population of cells that are adequate in health, homogeneity and quantity for subcellular or other investigations.

Whilst this book does not obviate the need to consult literature, its content is largely non-specialist and appropriate for anyone venturing into an unfamiliar field entailing, for example, the study of particular types of blood cell, secretory cell or unicellular parasite. Cancer-cell studies get scant attention here, but will feature in the 7th Forum and in a resulting book (Vol. 11 of this series; 1981).

The evolution of this series.— In case readers or librarians find difficulty in discerning a 'pattern' (cf. the listing opposite the title page), an explanation may help. Like the Wolfson Bioanalytical Centre itself (an R & D laboratory), the meetings and the non-ephemeral books which they generate are distinctive in fostering good methodology, largely in two areas hitherto apt for the term 'biochemistry', albeit far apart. That one area concerns cells is evident from the present book and its companion, *Plant Organelles* (Vol. 9). On the other hand, Vols. 5 and 7 have dealt with the assay of biological fluids for drugs and hormones, i.e. small molecules. With intended extension of the latter area to environmental contaminants (Vol. 10), the term 'biochemistry' becomes inappropriate: the new designation for this sub-series will be *METHODOLOGICAL SURVEYS (A): Trace-Organic Analysis.* The present book inaugurates the sub-series *METHODOLOGICAL SURVEYS (B): Biochemistry,* and will still lean towards 'cellular' rather than 'molecular' aspects (cf. *Biochemical Journal* sections). Whilst volume nos. will remain consecutive, a librarian may prefer to specify only *A* or *B* when building up a run (retrospectively if desired). Editorial gratification is expressed concerning the evolving alliance with Ellis Horwood Limited, as initiated with Vol. 6.

Acknowledgements.— These are multiple and heart-felt, not least to Mrs. R. Sarker for skilled sub-editing help as well as typing. The authors were mostly prompt, in spite of other pressures, and invariably good-natured. Dr. D.J. Morré was a notable ally in planning and running the Forum, along with other Honorary Advisers including Drs. G.B Cline, R. Coleman, H. Glaumann and G. Siebert. The Forum was co-sponsored by the European Cell Biology Organization, and received financial aid from company sources: Beckman-RIIC, Ciba Research, MSE, Nyegaard, and Pharmacia. Other publications have sometimes been drawn on, by courtesy of the Editor concerned, as is mentioned where applicable — particularly in certain Table headings and Fig. legends.

COMMENTS ON NOMENCLATURE

The terminology for cell types in liver, and for dispersions of cells that may or may not be homogeneous, shows inter-laboratory differences which are reflected in this book insofar as authors' preferences have been respected. However, as a contribution to international debate some thoughts are now set down.

Cells within or from liver

At the risk of being in a minority, the Editor and his colleague Dr. R.H. Hinton are inclined to favour stretching the term *parenchyma* to include sinusoidal cells, as in Scheme 1. This scheme

Scheme 1. Nomenclature preferred by the Editor and Dr. R.H. Hinton for cell types observed in liver *or (italics) isolated from liver*, in accord respectively with A.W Ham [1] and E.R. Weibel and co-authors [2]. Note that in some laboratories, especially biochemical, the term *parenchymal (or parenchymatous)* is restricted to hepatocytes, and sinusoidal cells are termed *non-parenchymal*.

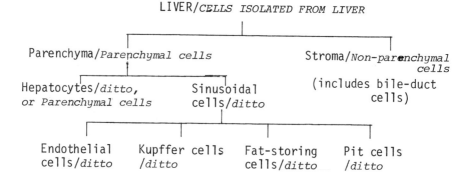

LIVER/*CELLS ISOLATED FROM LIVER*

Parenchyma/*Parenchymal cells*   Stroma/*Non-parenchymal cells*

Hepatocytes/*ditto, or Parenchymal cells*   Sinusoidal cells/*ditto*   (includes bile-duct cells)

Endothelial cells/*ditto*   Kupffer cells/*ditto*   Fat-storing cells/*ditto*   Pit cells/*ditto* gives terms encountered in histology and *(italicized)* in some studies with isolated cells. Elsewhere in this book, however, there is implicit adherence to a simpler scheme and, moreover, designation of sinusoidal cells as *non-parenchymal*, as is noted editorially. "Nonparenchymal cells in the liver of a young adult rat consist of sinusoidal cells, haemopoietic cells, bile duct cells, connective tissue cells and blood vessel wall cells" [3]. With able initiative from D.L. Knook, debate continues in the columns of *Bull. Kupffer Cell Foundation*.

1. Ham, A.W. (1965) *Histology* (5th edn), Lippincott, Philadelphia.
2. Blouin, A., *et al.* (1977) *J. Cell Biol.* *72*, 441-455.
3. Knook, E.L. & Sleyster, E.Ch. (1977) —*Ref.* 1 *in* #B-2, *this vol.*

Designation of cell preparations in relation to homogeneity

Guidance on nomenclature was sought by the Editor from Honorary Advisers before the Forum, and from Forum participants in a debate chaired and ably summarized by D.R. Headon. Amongst those who contributed helpful views were R.J. Hay, F. Leighton, G. Siebert, J.F. Tait, and notably C.N.A. Trotman— who advocates the term *Isolated cells*, with no cell-type specified, for suspensions of single cells derived from a tissue. Whilst he is in good company, the risk of confusion with purified cells inclines this Editor to favour the nomenclature in Scheme 2, where the first category has alternatives that are acceptable to Trotman (as is *Cell suspension*). Either *Purified* or *Enriched hepatocytes* could, however, be ranked as *Isolated hepatocytes* if desired.

Near-equivalent terms; starting material	*Suspended cells* - as already present in suspension cultures or fluid 'tissues'. *(Dispersed cells)* esp. from *Dissociated cells* a tissue.
From a secondary step (→ fractions)	*Separated cells* - from a fractionation acc. to (e.g.) size or density; individual cell type *not* implied.
Selectivity perhaps by mere digestion; maybe no secondary step	*Purified cells* - attainment of a homogeneous cell population for the desired aim; e.g. *purified hepatocytes*. An *Enriched cell preparation* (e.g. of *hepatocytes*) does not imply assured purity.

Scheme 2. Nomenclature which does not entail use of the term *isolated*. Note that *separated* is an operational term, and that ......*cytes* (e.g. *astrocytes*) implies proved/probable near-homogeneity. The starting material *(cell suspension)* may be sub-categorized as *dissociated*.

Wolfson Bioanalytical Centre
University of Surrey
Guildford

9 April, 1979

Eric Reid

Table of Contents

List of Authors

The post-codes complement the address listings on individual title pages.
Where the page no. is higher than 213, the item is a 'Note' in section NC.

R. Balázs – pp. 225–226
London WC1N 2NS, U.K.

W. S. Bont – pp. 53–66
1066 CX Amsterdam, Netherlands.

J. Cohen – *as for* R. Balázs

N. Crawford – pp. 191–202
London WC2A 3PN, U.K.

M. J. Crumpton – pp. 203–211
Mill Hill, London NW7 1AA, U.K.

J. E. de Vries – *as for* W. S. Bont

M. Edwards – *as for* H. A. Krebs

E. Eriksson – *as for* G. Johansson

W. H. Evans – pp. 7–13
Mill Hill, London NW7 1AA, U.K.

J. M. Greene – pp. 222–223
Cork, Eire.

R. R. L. Guillard – pp. 171–175
Woods Hole, MA. 02540, U.S.A.

W. Haase – *as for* I. Schulz

K. Hannig – pp. 91–103
D–8033 Martinsried b. München,
W. Germany.

R. J. Hay – pp. 143–160
and 227–231
Rockville, MD 20852, U.S.A.

D. R. Headon – pp. 222–223 (*see
footnote*)
Galway, Eire.

H.-G. Heidrich – *as for* K. Hannig

K. Heil – *as for* I. Schulz

M. Hirtenstein – pp. 67–80
Box 175, S-751 04 Uppsala, Sweden.

J. Hsiao – *as for* F. Ungar

Y. Ito – pp. 217–219
Bethesda, MD. U.S.A.

G. Johansson – pp. 81–90
P.O. Box 740, 5-220 07 Lund,
Sweden.

L. Kågedal – *as for* M. Hirtenstein

D. L. Knook — pp. 47–52
Rijswijk 2H, Netherlands

H. A. Krebs — pp. 1–6
Oxford OX2 6HE, U.K.

S. M. Lanham — pp. 177–190
London WC1E 7HT, U.K.

P. Lund — *as for* H. A. Krebs

J. P. Luzio — pp. 161–170
Cambridge CB2 2QR, U.K.

S. Milutinović — *as for* I. Schulz

M. J. Owen — *as for* M. J. Crumpton

H. Pertoft — pp. 67–80
Box 575, S-751 23 Uppsala, Sweden.

C. A. Price — pp. 171–175
Piscataway, NJ 08854, U.S.A.

R. G. Price — pp. 105–110
London W8 7AH, U.K.

G. Raydt — pp. 137–142
D-8000 München 2, W. Germany.

E. M. Reardon — *as for* C. A. Price

A. M. Robinson — *as for* D. H.
Williamson

G. Rumrich — *as for* I. Schulz

E. Schnell — *as for* G. Siebert

I. Schulz — pp. 127–135
D-6000 Frankfurt/Main, W. Germany

P. O. Seglen — pp. 25–46
Oslo 3, Norway

G. Siebert — p. 221
D-7000 Stuttgart 70, W. Germany.

E. Ch. Sleyster — *as for* D. L. Knook

G. Sturm — *as for* G. Siebert

I. A. Sutherland — pp. 217–219
Mill Hill, London NW7 1AA, U.K.

D. Terreros — *as for* I. Schulz

C. N. A. Trotman — pp. 111–126
Newcastle-upon-Tyne NE1 7RU, U.K.

F. Ungar — pp. 222–223
Minneapolis, MN 55455, U.S.A.

T. J. C. van Berkel — pp. 15–24
P.O. Box 1738, 3000 DR Rotterdam,
Netherlands.

D. H. Williamson — pp. 223–224
Oxford OX2 6HE, U.K.

M. H. Wisher — *as for* W. H. Evans

P. L. Woodhams — *as for* R. Balázs

#A Features of Isolated Liver Cells

#A-1
CRITERIA OF METABOLIC COMPETENCE OF ISOLATED HEPATOCYTES

HANS A. KREBS, PATRICIA LUND and MELFYN EDWARDS
Metabolic Research Laboratory,
Nuffield Department of Clinical Medicine
Radcliffe Infirmary, Oxford.

No single test supplies full information about the metabolic competence of isolated hepatocytes. Exclusion of trypan blue does not always parallel metabolic competence. The trypan blue exclusion test is useful only in a limited way in that if a high percentage of the cells do not exclude the dye the suspension is unsatisfactory. Additional tests are the assay of the total adenine nucleotides, the measurement of the rate of O_2 uptake, of gluconeogenesis from lactate and of the synthesis of urea. The response to glucagon provides information on the integrity of the outer membrane. When a metabolic process is used as a criterion, it must be measured under optimum conditions.

We reiterate that there is no simple single criterion of metabolic competence. We stress this point because many authors offer in their publications the comment that their hepatocytes appeared normal under the light microscope and that over 80% or 90% excluded trypan blue. One gains the impression that this satisfies them that their cells are competent (or as some people like to call it, viable). The trypan blue exclusion test is useful but only in a very limited way. If a high proportion of cells do not exclude trypan blue the suspension is unsatisfactory. Thus the test is a quick and simple indicator of inadequate cells. On the other hand dye exclusion is not a guarantee of metabolic competence. The following examples illustrate this statement.

ASSESSMENT OF METABOLIC COMPETENCE

Hepatocytes from 48 h starved rats were prepared by our modification [1] of the method of Berry & Friend [2]. A small sample (0.2 ml) of cell suspension was mixed with 0.1 ml of trypan blue (0.2% in 0.9% NaCl). At least 200 cells were counted in a haemacytometer. Table 1 shows that aerobic or anaerobic storage of cell suspensions for 3 h at 22° had relatively little effect on

trypan blue exclusion compared with the effect, especially of
anaerobic storage, on the subsequent capacity of the cells to
synthesize glucose from lactate. Ketogenesis from butyrate,
being a mitochondrial process, was resistant to the damaging
effects of storage.

Striking differences between changes in dye exclusion and
changes in gluconeogenic rate were found on addition of surface-
active agents. As illustrated in Table 2, 0.025% Triton X-100
inhibited gluconeogenesis completely but increased the proportion of cells taken up the dye by only 16%. In the case of
Cetavlon there was a rough parallelism between dye exclusion and
biosynthetic capacity.

Butanol and n-pentanol (amyl alcohol), which are also surface-
active agents, at certain low concentrations markedly inhibited
gluconeogenesis without affecting dye exclusion (Table 3). On
the other hand toluene decreased dye exclusion near-parallel
with the rate of gluconeogenesis. Again ketogenesis from butyrate was less affected, which confirms that mitochondrial processes are less sensitive to surface-active agents than cytosolic processes. These agents are chemically relatively inert,
and they are taken to interfere with cell activity on account of
either their lipid solubility or their adsorption at interfaces
[3]. In the present context the precise mechanism of their
action is unimportant as it is the object of the exercise to
demonstrate the limited value of dye exclusion tests. An alternative to the trypan blue test (which stains nuclei) is the
addition of succinate together with iodonitrotetrazolium salt
(INT) which stains mitochondria within a few minutes. INT is
reduced within the mitochondria to a red formazan by the flavo-
protein of succinate dehydrogenase. This is essentially a test
for the exclusion of succinate which normally does not enter
hepatocytes. To 0.2 ml cell suspension are added 0.04 ml of a
solution containing 13 mM INT, 60 mM succinate and 80 mM NaCl.
In general the two methods give the same results except when
the suspensions are treated with both dyes simultaneously. Under
these conditions about 5% stain with trypan blue only, possibly
because the mitochondria have leaked out of the cell through
major ruptures of the plasma membrane (p.m.).

DISCUSSION

The upshot is that the results of dye exclusion tests do not
necessarily run parallel with metabolic impairment. Mammalian
cells are very complex systems with numerous components, some
of which are more sensitive to damaging circumstances than
others. Hence there is no short-cut, by way of a simple dye
test, to the assessment of metabolic competence.

Table 1. Effect of storage of hepatocytes on trypan blue exclusion and on subsequent capacity to form glucose and ketone bodies.
Substrates and initial concns.: 10 mM lactate, 10 mM NH_4Cl,
2 mM ornithine, and 1 mM oleate (for glucose synthesis) or 10 mM
butyrate (for ketone body synthesis). Hepatocyte suspensions
were stored without albumin. Albumin (25 mg/ml) was present
during incubation with substrates for 60 min. The values are
the means of 4 experiments, with % change given in parentheses ().

Cell material	Trypan blue exclusion,%	Gluconeo-genesis, μmol/min/g	Ketogenesis, μmol/min/g
Fresh	86	1.49	3.07
Stored 3 h; 22°; 5% CO_2 in O_2	71 (-17%)	0.83 (-44%)	2.89 (-6%)
Stored 3 h; 22°; 5% CO_2 in N_2	68 (-21%)	0.50 (-66%)	2.38 (-22%)

Table 2. Effects of surface-active agents (detergents) on trypan blue exclusion and gluconeogenesis from lactate.
Substrates and initial concns. :   10 mM lactate, 1 mM pyruvate, 2 mM lysine, 1 mM oleate and 0.1 mM dibutyryl cyclic AMP.
Incubation for 30 min at 37°. Dye exclusion was determined
after incubation. The % change is given in parentheses.

Agent added	Final concentration,%	Trypan blue exclusion,%	Gluconeo-genesis, μmol/min/g
None	-	85	2.08
Triton X-100	0.00625	78 (-8%)	0.90 (-57%)
	0.0125	80 (-6%)	0.39 (-81%)
	0.025	72 (-16%)	0.05 (-98%)
	0.05	2 (-98%)	0.04 (-98%)
Cetavlon	0.00625	85 (0%)	1.68 (-19%)
	0.0125	89 (0%)	1.51 (-27%)
	0.025	28 (-67%)	0.93 (-55%)
	0.05	0 (-100%)	0.20 (-91%)

   The intactness of the cell depends above all on the permeability characteristics of the p.m. and of internal membranes,
especially the inner mitochondrial membrane, because these membranes are responsible for maintaining the normal and specific
intracellular and intraorganelle environment necessary for the
physiological operation of the enzymes. If the membranes do not

Table 3. Effects of surface active agents (higher alcohols and toluene) on trypan blue exclusion and on gluconeogenesis and ketogenesis.
Substrates and initial concns. : 10 mM lactate, 2 mM lysine, 1 mM oleate, and 0.1 mM dibutyryl cyclic AMP (for glucose synthesis) or 10 mM butyrate (for ketone body synthesis). Incubation for 60 min at 37°. Dye exclusion was determined after incubation. The values are expressed as % of control.

Agent added	Final concentration,%	Dye exclusion	Gluconeogenesis	Dye exclusion	Ketogenesis
n-Butanol	0.25	98	83	100	96
	0.5	87	47	100	91
	1.0	64	13		
n-Pentanol	0.25	80	2	84	41
	0.5	44	1	47	5
Toluene	0.25	86	88	100	94
	0.5	58	50	69	85
	1.0	2	2		

function normally, especially when they become leaky, the cells discharge low molecular constituents and even larger molecules such as enzymes; this upsets the internal environment.

The implication is that a series of criteria such as those listed below must be used for a reliable assessment of the functional integrity of the cells.

Appropriate criteria

(1) Loss of refraction, as revealed by phase contrast microscopy, indicates unduly swollen and presumably damaged cells.

(2) Staining with trypan blue or other dyes, as already mentioned, is a rough test in as much as major staining is definitely indicative of major damage.

(3) The assay of the total adenine nucleotides provides information on whether the key agents in energy transformations are maintained. Normal values imply that cell respiration and oxidative phosphorylation function normally.

(4) The rate of oxygen uptake is a useful quantitative measurement of the degree of integrity, especially when the time course is followed manometrically.

(5) Rates of gluconeogenesis from lactate or urea synthesis from ammonium chloride under optimal conditions are criteria of the adequacy of energy supply and energy utilization. Urea synthe-

sis is somewhat less exacting than glucose synthesis because it
is less dependent on physiological adenine nucleotide concentrations [4].

(6) A further test directed especially towards the check of the
integrity of the outer membrane is the response to glucagon.
The peptide hormones are taken to exert their effects through
receptor sites on the p.m. It would not be surprising if receptors were damaged by the crude collagenase at the perfusion
stage, but experience indicates that the receptors still function though it is not certain how much of their original capacity
is retained. Tests with glucagon on the stability of glycogen
show that very low glucagon concentrations down to 10^{-9} M accelerate glycogenolysis. The responsiveness of gluconeogenesis to
glucagon is an alternative check of the integrity of the p.m.

When testing for metabolic competence it is essential to carry
out the tests under optimum conditions where rates are at their
maximum. It is not good enough to demonstrate merely that gluconeogenesis from lactate occurs. Unless the capacity is
stretched to the maximum, partial losses of activity would not
be detected. In the case of gluconeogenesis from lactate maximum rates occur in the presence of 10 mM lactate, 1 mM pyruvate,
2 mM lysine (which proved the best agent for restoring and maintaining the amino acids needed for the aspartate shuttle [5]),
1 mM oleate (source of energy) and 0.1 mM dibutyryl cyclic AMP.

For testing the maximum rate of urea synthesis it is necessary
to add not only NH_4Cl as a precursor, but also 1 mM ornithine as
a catalyst, 1 mM oleate as a source of energy and 2 mM lactate
as a precursor of intracellular aspartate required for the formation of argininosuccinate.

There are many other properties one could assess, e.g. the
ability to maintain normal intra/extracellular K^+ and Na^+ gradients. However gluconeogenesis appears to be rather insensitive
to alterations of the gradients: ouabain markedly decreases the
ion gradients and the intracellular concentration of alanine
without affecting the rate of gluconeogenesis from alanine [6].

A reason why gluconeogenesis from lactate is a particularly
exacting criterion is the fact that it is a complex process
which depends on the intactness of the energy supply, the intactness of the intracellular environment (especially the concentrations of the amino acids needed for the aspartate shuttle, the
intactness of ATP formation and of the intracellular ATP concentration, the intactness of regulatory mechanisms, the intactness
of the p.m. and the mitochondrial membranes). Thus one measurement, that of the rate of gluconeogenesis under maximum conditions, tests several key properties. It might be thought that
an alternative to gluconeogenesis is urea synthesis because both

involve the intactness of the p.m. and the mitochondrial membrane, but measurement of these rates in slices shows that gluconeogenesis is more easily damaged than urea synthesis. We sometimes do both tests as these are rather simple. In addition, we often measure O_2 uptake because this gives the time course if carried out manometrically. The rate of O_2 uptake should be almost constant during the period of the experiment. The suggestion is that anyone who proposes to work on isolated hepatocytes should satisfy himself that he can reproduce the maximum rates reported in the literature [see 7] before studying new aspects. Merely looking at cells — at their shape and at dye exclusion — is not good enough. Nor is reproducibility of data, because this does not reveal systematic errors. Thus a second or third sample is not a reliable reference standard. An acceptable reference standard for a new test is the isolated perfused intact liver.

Concluding comments

What has been said so far applies to rat hepatocytes. Chicken hepatocytes may require different criteria. We find that chicken liver cells are more resistant to unphysiological conditions. While rat hepatocytes usually lose about 10% of their metabolic capacity per hour, such losses are less in chicken hepatocytes. Furthermore in rat liver the rate of gluconeogenesis from lactate, as already stated, depends on the presence of oleate, cyclic AMP and lysine. In chicken liver these substances do not increase gluconeogenesis from lactate, which always gives high rates (about 6 µmol/min/g wet wt). This is more than twice the maximum rates observed in rat liver.

Finally, we would make a plea for uniform expression of metabolic rates. Some authors use µmol/min per g dry wt. or µmol/min per g wet wt. which are both acceptable. Others use as a reference the number of cells. This makes comparison of the rates of different authors difficult. It also complicates the comparison of different species. Avian hepatocytes are about 10 times smaller than mammalian hepatocytes.

References

1. Krebs, H.A., Cornell, N.W., Lund, P. & Hems, R. (1974) in *Regulation of Hepatic Metabolism,* Vol. 6 *(Alfred Benzon Symposium)* (Lundquist, F. & Tygstrup, N., eds.), Munksgaard, Copenhagen, pp. 726-750.
2. Berry, M.N. & Friend, D.S. (1969) *J. Cell Biol. 43,* 506-520.
3. Warburg, O. (1914) *Ergebnisse der Physiologie 14,* 253-337.
4. Tager, J.M., Zuurendonk, P.F. & Akerboom T.P.M. (1973) in *Inborn Errors of Metabolism* (Hommes, F.A. & van den Berg, eds.), Academic Press, London, pp. 177-197.
5. Cornell, N.W., Lund, P. & Krebs, H.A. (1974) *Biochem. J. 142,* 327-337.
6. Sainsbury, G.M. (1978) *D. Phil Thesis,* Oxford University.
7. Krebs, H.A., Lund, P. & Stubbs, M. (1976) in *Gluconeogenesis* (Hanson, R.W. & Mehlman, M.A., eds.), Wiley, Chichester, pp. 269-291.

#A-2
PROPERTIES OF THE CELL SURFACE OF ISOLATED HEPATOCYTES

W. HOWARD EVANS and MARTIN H. WISHER
National Institute for Medical Research,
Mill Hill, London NW7, U.K.

The effects on the hepatocyte plasma membrane (p.m.) of enzymic dissociation of liver tissue is evaluated from the viewpoint of damage to the intercellular junctions and to p.m. ectoenzymes. The consequences to the hepatocyte's functional polarity of removal from an organized epithelium are discussed.

The process of dissociating tissues and organs into the constituent cells is somewhat traumatic. In liver, a range of methods for tissue disaggregation have been explored to obtain primary monolayer cultures [1], and current methods now rely almost exclusively on enzymic dissociation using collagenase [2, 3; *see also* P.O. Seglen, *this volume*]. In the present account, the consequences to the p.m. of liver tissue dissociation are discussed from three aspects. First, the effects of the dissociation regime on the intercellular junctions are discussed in the context of current knowledge of junctional morphology and their molecular architecture. Second, damage to p.m. ectoenzymes is assessed by examining the recovery from isolated cells of p.m. 5'-nucleotidase and nucleotide pyrophosphatase. Finally, the effects of removal of hepatocytes, in rendering the cells free from an organized epithelium that separates blood and bile compartments, is examined from the viewpoint of whether biochemically characterized p.m. subfractions are still obtainable.

EXPERIMENTAL

Tissue dissociation and cell separation

Rat livers were dissociated *in situ* by perfusion through the portal vein with a Ca^{2+}-free Hank's medium pH 7.4 at 37°, equilibrated in O_2/CO_2 (19 : 1 by vol.) using a Miller-type re-circulating perfusion apparatus [4]. After flushing out blood from the organ with 50 ml of Hank's medium, perfusion was continued at 10-25 ml/min with 100 ml Hank's medium containing 50 mg collagenase (Sigma Type 1 prepared from *Clostridium histolyticum*) and

5 mg soya bean trypsin inhibitor (Sigma). After 15-25 min, the
liver was removed, agitated with a spatula and filtered through
cheesecloth and a nylon mesh (6 μm pore size; Henry Simon Ltd.,
Cheadle Heath, Stockport, U.K.). Cells were separated at unit
gravity, or by low-speed centrifugation (50 g for 2 min) or by
centrifugation in Ficoll gradients (Pharmacia) in which non-
parenchymal cells were collected at the 0-15% interface, viable
hepatocytes at the 15-30% interface and non-viable hepatocytes
at the bottom of the tube. Viability was assessed by Trypan blue
exclusion and by electron microscopy of fixed and stained cells.

Enzyme assays and subcellular fractionation

5'-Nucleotidase was determined spectrophotometrically using 5'-
AMP as substrate in the presence of an adenosine deaminase, and
nucleotide pyrophosphatase was measured as alkaline phosphodiesterase II using thymidine 5'-monophospho-p-nitrophenyl ester as
substrate. Isolated hepatocytes, suspended in 0.25 M sucrose at
4°, were homogenized in a Dounce homogenizer using a tight-fitting
pestle (radial clearance 0.076 mm; Blaessig Glass Co., Rochester,
New York, U.S.A.) until >90% of the cells were disrupted as observed by phase-contrast microscopy. Then p.m. subfractions were
prepared and characterized - enzymically, morphologically and
chemically - as described [5].

Electron microscopy

Samples were fixed for 1 h at 4° in a mixture (2 : 1 by vol.) of
1% OsO_4 and 2.5% glutaraldehyde dissolved in sodium cacodylate
buffer, pH 7.4 [6]. The pellet was 'post-fixed' for 15 min in
0.25% uranyl acetate dissolved in 0.1 M veronal acetate buffer,
pH 6.2 and embedded in Epikote 812. Thin sections were stained
in uranyl acetate and lead citrate and observed in a Philips
EM-300 microscope.

RESULTS AND DISCUSSION

Effects of tissue dissociation on intercellular junctions

Cells in tissues of higher organisms interact through three major
types of junctional complexes, viz. desmosomes, tight junctions
and gap (communicating) junctions, and these have to be severed
before cell separation can take place.

Desmosomes are discrete zones of adhesion between cells; hemi-
desmosomes serve as anchoring sites to the underlying matrix of
connective tissue. Strictly, desmosomes are mainly specializations of the inner and outer surfaces of the p.m. Characteristic
tonofilaments are attached to the inner surface of the p.m., and
a dense plaque of unknown composition fills the 300 Å gap between
the outer aspects of the apposed p.m.'s. The plaque material is
readily dissociated by the collagenase-containing media, and in
dissociated hepatocytes membrane thickening that probably

corresponds to desmosome 'halves' can be recognized at the point
of rupture (Fig. 1a). Desmosome 'halves' are probably internalized and hydrolyzed [7]. Cow nose epidermis desmosomes have
been isolated and analyzed [8].

Tight junctions arc intcrluminal pcrmcability scals forming
belt-like structures completely surrounding cells. Freeze-
fracturing shows that they consist of fibrils of closely-packed
intramembranous particles sealing off the extracellular space,
and zippering the two cells together [9]. It is difficult to
generalize about the ease of dissociation of the p.m. region con-
stituting the tight junction in various epithelia, in view of the
variation in tissue permeability [10]. In the liver, the tight
junction seals the bile canaliculus, thus forming a barrier preventing bile components from diffusing into the intercellular
space that lead to the blood sinusoids. In extracellular choles-
tasis, the build-up of pressure of bile secreted into the canaliculi can lead to the opening of the tight junction barrier, so
allowing the bile a direct pathway to the hepatic blood sinusoids;
assay of bile products in blood constitutes a clinical test for
this syndrome. Thus, it may be reasoned that at least in liver
tissue the p.m.'s comprising the tight junction can be torn apart
under pressure, and that gentle mechanical treatment as always
applied in liver disaggregation using collagenases may also
suffice to separate the apposing membranes. Observation of the
surface regions corresponding to the bile canaliculus and the
adjacent p.m.'s suggest that tight junctions are severed when
siolated hepatocytes are prepared (Fig. 1b). Monolayer cultures
of hepatocytes reconstitute the tight junctions, so re-forming
bile canaliculi within 24 h of tissue dissociation [11]. There
are no reports of the isolation of tight junctions.

The gap junction is the p.m. specialization that mediates
direct intercellular transfer of ions and small molecules. They
consist of zones of tightly interacting intramembranous particles
of the contiguous p.m.'s that are believed to construct the
intercellular channel. However, a 'gap' still remains between
the cells, so maintaining an intercellular pathway along which
can pass heavy atom tracers, e.g. lanthanum.   So tight is the
interaction of the gap junction particles that separation of the
contiguous membranes rarely occurs. Complete gap junctions are
observed on the surfaces of isolated hepatocytes (Fig. 1c) .
Indeed, when p.m. fractions are prepared from isolated hepatocytes, gap junctions are recognized (Fig. 1d) that correspond to
those used to identify p.m. material prepared from liver [2]. The
contributing halves of gap junctions interact so tightly that
they are degraded following internalization as complete entities
[13]. The retention of the complete gap junction structure on
the surface of a hepatocyte must result in the removal of a
corresponding region of p.m. of the neighbouring hepatocyte.

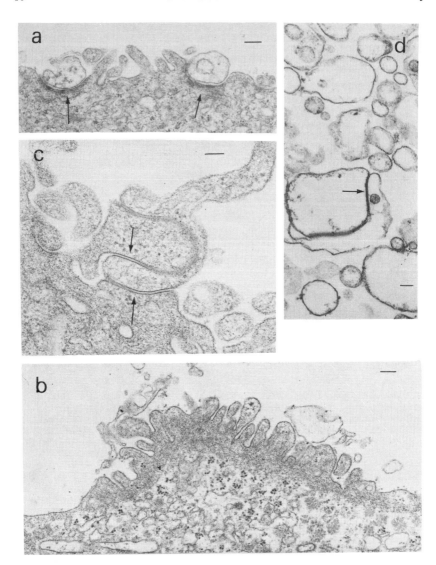

Fig. 1. Intercellular junctions and surface specializations
present on isolated hepatocytes. The bar represents ∿ 0.2 μm or,
in 1d, ∿0.1 μm.

1a: Remnants *(arrows)* on hepatocyte surface of two desmosomes.

1b: Microvillar area that probably corresponds to the bile canalicular region; tight junctions on each side appear to have
been ruptured.

1c: Two intact gap junctions *(arrows)* and attached cytoplasm torn
from an adjacent hepatocyte.

1d: Gap junction *(arrow)* present in a p.m. fraction prepared from
isolated hepatocytes.

Although re-sealing of the p.m. may occur, this rupture of the membrane must also lead to the loss of intracellular metabolites, and thus contribute to the number of non-viable hepatocytes produced when dissociating tissues. This must be especially so when large gap junction areas are present. Although desmosomes and tight junctions re-appear when hepatocytes are maintained for 24 h [11], re-formation of gap junctions and thus the establishment of metabolic co-operation between hepatocytes is further delayed. Gap junctions have been prepared from liver tissue and characterized [14, 15].

Effects of enzymic dissociation on hepatocyte cell-surface enzymes

Those p.m. enzymes that are exposed to the extracellular matrix (ectoenzymes) are ideally positioned for damage by that proteolytic cocktail termed collagenase used to dissociate tissues. Some effects of collagenase treatment became apparent when two liver p.m. ectoenzymes were used as markers in the subcellular fractionation of isolated hepatocytes. When the increases in the specific activities of p.m. marker enzyme relative to the liver tissue or hepatocyte cell homogenates were determined, the results were more comparable when a soya-bean trypsin inhibitor was included with the collagenase used in tissue dissociation (Table 1). These results emphasize how proteolytic damage to the cell surface may be minimized during tissue dissociation. However, the effects of collagenase on cell surface components are complex, for exposure of p.m. fractions to collagenases failed to modify significantly the above two enzyme activities. Thus, the major result was an approximately two-fold increase in the recovery in p.m. subfractions of the two ectoenzymes examined.

Since leakage of nucleotides and other metabolites may ensue following surface damage especially at points at which intercellular contact occurred, it is important that the p.m. enzymes discussed above remain active, for alone or in combination they can provide a mechanism to ensure the conversion of extracellular nucleotides into nucleosides that can be transported across the p.m. into the cytosol, as recently shown with cardiac p.m. 5'-nucleotidase [16]. It is likely that viable hepatocytes can repair the damage to surface components over a period of hours, since membrane synthesis and turnover proceed, as shown in other cell types [7].

Effect of enzymic disociation on the functional polarity of hepatocytes

In liver tissue, the organization of hepatocytes ensures that there is transepithelial movement of metabolites initially taken up by the blood-sinusoidal p.m. and then, after modification, released into bile, and that secretory products are directed to the correct surface domain. A fundamental question pertaining

Table 1. Comparison of 5'-nucleotidase and nucleotide pyrophosphatase activities in p.m. prepared from liver tissue and isolated hepatocytes.
For details of preparation, see [5]. 'Light' and 'heavy' denote densities in sucrose gradients of 1.12-1.14 and 1.16-1.18 g cm³, respectively. The figures in parentheses are the recoveries in the p.m. subfractions of homogenate enzyme activities.

Enzyme	Subfraction of p.m.	Specific activity relative to the homogenate		
		Hepatocytes prepared :		Liver tissue
		without inhibitor	with soya-bean inhibitor present	
5'-Nucleotidase	Nuclear 'light'	11.3 (0.95)	37.3 (1.80)	74.6 (4.8)
	Microsomal 'light'	14.2 (6.5)	24.4 (12.1)	41.7 (15.0)
	Nuclear 'heavy'	8.3 (0.47)	18.6 (0.50)	22.6 (0.80)
Nucleotide pyrophosphatase	Nuclear 'light'	20.3 (1.71)	41.0 (2.45)	110 (7.1)
	Microsomal 'light'	25.0 (11.5)	32.7 (17.5)	45.9 (21.5)
	Nuclear 'heavy'	10.9 (0.65)	19.7 (0.78)	28.0 (0.97)

to isolated hepatocyte physiology concerns the degree of retention of the functionally distinct blood sinusoidal and bile canalicular domains. With isolated hepatocytes, p.m. subfractions that corresponded broadly on the basis of enzymic and morphological properties to those obtained from liver tissue have been prepared and characterized [5, 18]. The results reinforced the expectation that polarity of organization studied at the p.m. level is but a reflection of intracellular functional compartmentation that is not substantially modified when viable isolated hepatocytes are prepared. Thus, the intracellular organization ensuring, for example, that albumin and bile salts are directed to and released at different regions of the cell surface is a result of the same intracellular compartmentation that leads to the development in monolayer hepatocyte cultures of bile canaliculi surrounded by tight junctions [10].

Clearly, a study of the surface properties and of p.m. events in metabolically competent isolated hepatocytes can provide much information about the processes that help overcome the trauma undergone by epithelial cells after they are enzymically and mechanically separated from one another, followed by cellular re-assembly to form 'mini'-tissues.

References

1. Waymouth, C. (1974) *In vitro 10,* 97-111.
2. Howard, R.B., Christensen, A.K., Gibbs, F.A. & Pesch, L.A. (1967) *J. Cell Biol. 35,* 675-684.
3. Berry, M.N. & Friend, D.S. (1969) *J. Cell Biol. 43,* 506-520.
4. Miller, L.L., Bly, C.G., Watson, M.L. & Bale, L.F. (1951) *J. Exp. Med. 94,* 431-453.
5. Wisher, M.H. & Evans, W.H. (1977) *Biochem. J. 164,* 415-422.
6. Hirsch, J.G. & Fedorko, M.E. (1968) *J. Cell Biol. 38,* 615-627.
7. Overton, J. (1968) *J. Exp. Zool. 168,* 203-214.
8. Skerrow, C.J. & Matolsty, A.G. (1974) *J. Cell Biol. 63,* 515-530.
9. Staehelin, L. (1973) *J. Cell Sci. 13,* 763-786.
10. Diamond, J.M. (1974) *Fed. Proc. 33,* 2220-2224.
11. Wanson, J.C., Drochmans, P., Mosselmans, R. & Ronveaux, M.F. (1977) *J. Cell Biol. 74,* 858-877.
12. Evans, W.H. (1970) *Biochem. J. 116,* 833-842.
13. Larsen, W.J. (1977) *Tissue and Cell 9,* 373-394.
14. Goodenough, D.A. (1976) *J. Cell Biol. 68,* 220-231.
15. Culvenor, J.G. & Evans, W.H. (1977) *Biochem. J. 168,* 475-481.
16. Frick, G.P. & Lowenstein, J.M. (1977) *J. Biol. Chem. 253,* 1240-1244.
17. El-Allaway, R.M. & Gliemann, J. (1972) *Biochim. Biophys. Acta 273,* 97-109.
18. Wisher, M.H. & Evans, W.H. (1977) in *Membrane Alterations as Basis of Liver Injury* (Popper, H., Bianchi, L. & Reutter, W., eds.), M.T.P., Lancaster, England, pp. 127-141.

#A-3

ENZYMOLOGICAL IDENTIFICATION OF CELL TYPES FROM LIVER

THEO. J. C. VAN BERKEL
Department of Biochemistry I,
Erasmus University,;
Rotterdam, The Netherlands.

The availability of procedures whereby parenchymal and 'non-parenchymal' cells (PC, NPC)* can be isolated from liver intact and pure enables the different cell types from liver to be bio-chemically characterized. The purity and integrity of the cell preparations can be tested in one assay by assaying simultaneously for the cytoplasmic enzyme L-type pyruvate kinase (PK), which is PC-located, and for M_2-type PK, which is NPC-located. Determination of different enzyme activities bound to one cell organelle offers the possibility to test the functional requirements of the studied cell organelle in the different cell types. The distribution of (iso)enzymes involved in the glycolytic-gluconeogenic pathway indicates that gluconeogenesis is restricted to PC. Mitochondrial enzymes show relatively low activities in NPC ($\sim$15%) compared with PC. NPC are relatively enriched in lysosomal enzymes (especially cathepsin D and acid lipase) and peroxidase: seemingly PC possess the set of lysosomal enzymes requisite for their autophagic role, while the NPC lysosomal enzymes can also exert a function in the breakdown of protein and lipid substrates from extracellular sources.

The mammalian liver consists primarily of hepatocytes (here termed parenchymal cells, PC). However, as many as 35% of the cells are sinusoidal ('non-parenchymal', NPC), representing 5-10% of the liver mass [1]. With liver and other organs having hetero-geneous cell populations it is difficult to assign unequivocally the various organ functions to the responsible cell type. The availability of cell separation procedures for liver allows the

---

*
EDITOR'S NOTE (cf. Preface).— The term 'parenchymal' as used by some workers, including E.R. Weibel [1, 19], encompasses not only hepatocytes but also sinusoidal cells; in the present paper the latter are termed 'non-parenchymal' ('NPC') and no consideration is given to other minor types [19], such as bile-duct cells.

study of the relative contribution of the different cell types to the total liver metabolism.

When working with cell organelles from liver it is important to recognize that these may originate from different cell types with different metabolic functions. Demonstration of heterogeneity, e.g. with isolated mitochondria as reviewed [2], may be due merely to different functional requirements of this organelle in a certain cell type. The biochemical studies discussed here allow one to assess how far such heterogeneities are attributable to NPC.

MATERIALS AND METHODS

Isolation of cells from liver

Various methods have been described for the isolation of PC and NPC. For PC the commonest method utilizes liver perfusion with 0.05% collagenase (according to Berry and Friend [3]). NPC isolation is performed by perfusing the liver with pronase [4], which specifically destroys PC while leaving NPC intact. The procedure used by our group (Scheme 1), utilizing both collagenase and pronase, furnishes from one liver both pure PC [5] and pure NPC, 'NPC$_1$' [6]. Furthermore, an 'NPC$_2$' preparation can be obtained by differential pelleting without the use of pronase [7], offering possible advantages in studying specific binding properties of NPC. An important advantage of the procedure described in Scheme 1 is that an enzyme distribution study can be undertaken within one liver, which permits the recovery of the enzymatic activity found in the different cells to be related to the value obtained with a whole rat-liver homogenate. Such a recovery calculation virtually rules out the possibility of artifacts associated with the assays or introduced by the cell isolation procedure. Further technical details are given elsewhere [5-8].

Test for cell purity and integrity

Although light microscopy and trypan blue exclusion can give an impression of the quality and purity of isolated cells, in biochemical studies the prevalence must be given to a biochemical criterion for purity and integrity. From our studies with the cytosol enzyme pyruvate kinase (PK) [9] we developed a simple assay (details in [8]) in which both the purity and integrity of each cell preparation can be tested. Rat liver contains two types of PK of which the L-type is PC-located and the M$_2$-type NPC-located [5, 9]. The difference in affinity of the two types for the substrate phosphoenolpyruvate (PEP) allows the application of a differential assay. The M$_2$-type PK exerts its maximal activity at 1 mM substrate and pH 8.0 and is not further activated by addition of the allosteric activator fructose 1,6-diphosphate (FDP). In contrast, under these conditions the activity of

Scheme 1. Isolation of
PC and NPC. Perfusion
and incubations were per-
formed in Hank's balanced
salt solution at 37°, and
filtration and centrifugation steps at 0°.

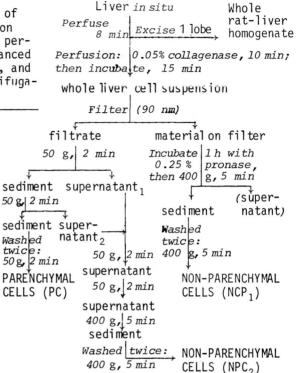

the L-type PK is sub-optimal and addition of FDP to the assay
gives an 11-fold stimulation.

Table 1 shows the activities obtained with the different cell
preparations. It can be calculated that 10% stimulation upon
addition of FDP to the PK assay with the NPC homogenates corresponds to a contamination of this preparation with PC of 1% (on
a cell number basis).

For cell integrity, in accordance with the views of H.A. Krebs
et al. *(this vol.)*, a criterion must be used which is relevant
to the particular study. It is because the present contribution
concerns enzyme localizations that the V_{max} activity of a cytosol
enzyme which does not adhere to protein-containing fragments
(unpublished results) is used. Intact PC must contain a pyruvate
kinase activity of at least 120 nmol/min/mg protein, while for
NPC this value is 35. The use of mitochondrial or other orga-
nelle-bound functions as criteria for integrity must be considered as inadequate [10, 11] in the context of cellular functions including cytosol involvement (cf. H.A. Krebs *et al.*, *this*
vol.).

Table 1. Purity and integrity test
on the isolated cell preparations
(Scheme 1) as determined by the distribution of L- and M_2-type pyruvate
kinase (PK). Activities are in
nmol/min/mg protein.

Source of homogenate	With 1 mM PEP	With 1 mM PEP + 0.5 mM FDP
Whole rat liver	20.4	133
PC	12.4	146
NPC_1	49.2	49.6
NPC_2	40.3	110.2

Measurement of enzyme activities

For the determinations of enzyme levels in PC and NPC, the usual
enzyme-assay requirements have to be fulfilled (activity proportional to amount of homogenate protein and linear with time of
measurement). Some special precautions have to be taken. It
was found with NPC preparations that, due to their high cathepsin
activity, storage at -20° could lead to inactivation of
cathepsin-sensitive enzymes [6]. In contrast, storage at -80°
preserved most activities completely. Moreover, homogenization
procedures for the cells have to be varied for each individual
enzyme activity: thus succinate dehydrogenase (SDH) is inactivated during sonication for 30 sec, whereas for measurements of
maximal activities of glutamate dehydrogenase (GDH) sonication
for 2 × 30 sec was found necessary [6].

    For our studies we developed the following protocol :—
(a) measurement of enzyme activity immediately after isolation
of the cells (in at least one experiment); (b) measurement of
the enzymic activity after storage at -80° before and after sonication of the homogenates; (c) checking of enzyme activity,
under optimal conditions, against time and protein concentration.
Possibly some of the discrepancies in the literature about the
activities of some NPC-located enzymes could be resolved by
using the above-mentioned criteria.

RESULTS AND DISCUSSION

Localization of cytoplasmic enzymes

For numerous enzymes involved in the glycolytic-gluconeogenic
pathway the distribution and isoenzymic form in PC and NPC have
been determined. Unfortunately in most of these studies no calculation has been performed to check the recovery of the isoenzymic activities in the different cell preparations as compared to
the whole rat liver homogenate. This means that generally it is
not excluded that upon cell isolation one isoenzymic form or
activity may be specifically lost. Although in the early studies
this omission largely reflected methodological problems, it must
be stressed that even in the most recent studies such a recovery
measurement is more the exception than the rule.

The early studies (1969) of Sapag-Hagar *et al.* [12] had
already indicated that glucokinase is present in PC cells, while
hexokinase is virtually restricted to NPC. Crisp & Pogson [13]
confirmed these results and showed further that NPC contained
one half of the specific activities of lactate dehydrogenase
(LDH), phosphoglycerate kinase and phosphofructokinase (PFK) as
compared to PC. Crisp & Pogson [13] and Van Berkel *et al.* [9]
found that the L-type PK is localized in PC and, with only about
half the specific activity, the M_2-type in NPC.

Recently work by Dunaway and co-workers [14] and Levin *et al.*
[15] indicated that PFK-L_2 is present in PC, while the other
hepatic PFK-L_1 is evidently found in NPC. This distribution of
isoenzymes (PK, PFK, glucokinase) indicates that most regulatory
forms (L-type PK, L_2-type PFK, glucokinase) are present in PC;
only these isoenzymic forms respond to hormones and dietary
fluctuations. The low or absent activity of the gluconeogenic
enzymes FDPase [13], pyruvate carboxylase [6] and glucose-6-
phosphatase (G-6-Pase) [13, 16] in NPC also indicates that it is
the parenchymal cells that respond to hormonal and dietary
changes and are able to regulate glucose homeostasis. Low FDPase
[33] activity in NPC was also found by histochemical techniques.
This means that gluconeogenesis is restricted to PC. Although
in a recent paper Wagle *et al.* [17] reported gluconeogenesis in
NPC at a rate 10-20% of that produced in hepatocytes, in the
absence of any biochemical criterion for purity it is unclear to
what extent their results can be explained by contamination of
the NPC with PC.

Besides enzymes linked to the gluconeogenic-glycolytic pathway, the activity distributions of superoxide dismutase [8],
glutathione reductase and glutathione peroxidase [18] have been
determined. The activity of the cytosolic forms of these enzymes
in NPC were about 50% of that of PC, indicating that these enzymes
have probably no specific function in NPC.

Fig. 1. Activities of the mitochondrial enzymes in NPC homogenates relative to those in PC homogenates. The activity ratio for cytochrome *c* oxidase has been taken as unity.

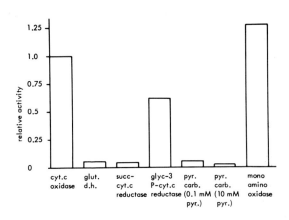

Mitochondria

The volume contribution of the mitochondria of NPC to the whole cell volume is only 15% of that of PC [19]. Recently we described the pattern of several mitochondrial enzymes in NPC [16], as summarized in Fig. 1. The specific activity of pyruvate carboxylase in NPC expressed as the percentage of that in PC was only 2%. This result strengthens the conclusion drawn from the distribution of cytosolic enzymes that the cells are not well equipped to perform gluconeogenesis. The relative high activity of the mitochondrial enzyme monoamine oxidase in NPC may indicate that these cell types are able to prevent toxic interference by biogenic amines with the respiratory chain.

The ratio of glycerol-3-P and succinate dehydrogenase activities differs between PC (0.01) and NPC (0.10). This is generally accepted as an indication that the glycerol-3-P shuttle, which is important for the transport of reduction equivalents from cytosol to mitochondria, is relatively more important in NPC than in PC. This may be related to the higher importance of glycolysis in these cell types. The lack of gluconeogenesis and the relatively low activity of pyruvate carboxylase limit the net dicarboxylic acid synthesis. Therefore the glycerol-3-P shuttle may compensate for a possibly moderate malate-aspartate shuttle activity in NPC.

Recently Fukushima *et al*. [20] showed that the glucagon-induced increase in serine-pyruvate aminotransferase activity mainly occurs in PC rather than NPC. These results therefore indicate that NPC lack a glucagon sensitivity at least for regulating this enzyme.

Fig. 2. Effect of pH
upon DAB peroxidase acti-
vity of PC and NPC homo-
genates. The activities
were measured in 0.1 M
Na-K-phosphate buffer.

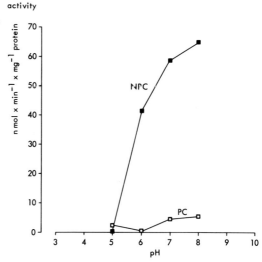

Peroxisomes

The peroxisomal enzyme
catalase is found in
NPC at about 15% of the activity of PC [5, 18], as recently confirmed [20]. In contrast, NPC have much higher peroxidase activity (Fig. 2), which rises considerably on increasing the pH
from 5 to 8; hence it can be used as a specific marker for NPC.
Histochemical studies (unpublished results) have shown that in
the NPC preparation only Kupffer cells and not endothelial cells
stain for peroxidase. In biochemical studies is therefore seems
possible to use peroxidase activity as a specific marker during
Kupffer cell isolation. Its function has been reviewed [21].

Plasma membranes (p.m.)

Since NPC are smaller than PC, p.m. are relatively enriched in
NPC [19]. Wincek *et al.* [16] showed that the p.m.-bound enzyme
adenyl cyclase is 5-fold and 5'-nucleotidase is 3-fold more
active in the NPC than in PC. The adenyl cyclase from NPC was
shown to be insensitive to glucagon stimulation [16], in full
accord with the lack of glucagon effects in these cell types.

NADPH oxidase, an enzyme which might function in the metabolic
activation which accompanies the phago(endo)cytotic process, is
about equally active in PC and NPC at neutral pH [22]. However,
by lowering of the pH to 5.5 only the NPC enzyme is activated,
thereby manifesting a property only found in phago(endo)cytic
cells. This enzyme is thought to play a role in the synthesis
of H_2O_2 within the phagocytic vacuole. As suggested earlier [23],
activation of this enzyme may well be the first signal by which
the uptake and degradation of particles by NPC from the bloodstream is initiated.

Endoplasmic reticulum

Crisp & Pogson [11] and Wincek *et al.* [16] found that the activity of the microsomal enzyme G-6-Pase was negligible in NPC preparations, being only 5% of that found in PC, in accord with the
lack of gluconeogenesis in these cell types. Cantrell & Bresnick
[24] found also low activity of benzpyrene hydroxylase (7.5%) in
NPC. Interestingly, in contrast to the low activities of the
afore-mentioned enzymes in NPC, haem oxygenase was found to
possess a 4-fold [25] or even 10-fold [26] higher activity in
NPC than in PC. This specific enrichment of haem oxygenase
might point to a special function of NPC in erythrocyte catabolism [27].

Lysosomes

Since the volume contribution of the lysosomes in NPC is 10-fold
higher than for PC [19], this cell organelle has been emphasized
in biochemical studies with NPC [28-30]. Several groups have
given, in the past or recently, distribution values for lysosomal
enzymes, with mostly comparable results. Our own data [28] have
indicated that NPC are not especially enriched in lysosomal
enzymes which hydrolyze the terminal carbohydrate moiety of
glycoproteins and glycolipids, or hydrolyze their sulphate and
phosphate esters. The most active enzymes in NPC are specialized
in protein (cathepsin D) and lipid (acid lipase) degradation
(Table 2). This led to the suggestion that NPC perhaps play a
specific role in lipoprotein catabolism. To test this possibility we have compared recently [31, 32] the capacities of PC and
NPC cells to degrade natural occurring lipoproteins. Fig. 3
shows that NPC possess an 8-fold higher capacity than PC cells
to degrade rat low-density lipoprotein (LDL). For rat high-
density lipoprotein (HDL) this factor is 5.5. It can further be
noticed that the curves follow saturation kinetics with half
maximal hydrolysis at about 12.5 µg/ml LDL apoprotein. On the
basis of this type of experiment it can be calculated that NPC
contribute 38% (HDL) or 46% (LDL) to the total rat liver capacity
for lipoprotein degradation. Recently the relative contribution
of rat PC and NPC to the hepatic uptake of HDL and LDL *in vivo*
was determined [7]. NPC isolated 6 h after intravenous injection
of iodinated HDL and LDL contained respectively 3.4 and 4.1 times
the amount of trichloroacetic acid-precipitable radioactivity
per mg cell protein as compared to PC. These results indicate
that NPC indeed play an important role in the hepatic uptake and
degradation of HDL and LDL *in vivo*. PC may, then, possess the
set of lysosomal enzymes to cope with their own autophagocytosis,
while the lysosomal enzymes from NPC can also exert a function
in the breakdown of protein and lipid substrates from extracellular sources.

Acknowledgements

Mr. J.K. Kruijt is thanked for continuing excellent technical assistance. Miss A.C. Hanson for her expert help in the preparation of the manuscript, and Dr. J.F. Koster, Dr. A. van Tol and Prof. Dr. W.C. Hülsmann for their interest and cooperation.

Table 2. Distribution of cathepsin D and acid lipase between PC and NPC. Activities are in nmol/min/mg protein and nmol/h/mg protein respectively.

Source of homogenate	Cathepsin D	Acid lipase
Whole rat liver	11.2	5.3
PC	6.9	2.9
NPC_1	41.6	29.8
Ratio NPC_1/PC	6	10

References

1. Weibel, E.R., Stäubli, W., Gnägi, H.R. & Hess, F.A. (1969) *J. Cell Biol. 42*, 68-91.
2. Geville, G.D. (1969) in *Citric Acid Cycle, Control and Compartmentation* (Lowenstein, J.M., ed.), Dekker, New York, pp. 1-136.
3. Berry, D.M. & Friend, D.S. (1969) *J. Cell Biol. 43*, 506-520.
4. Mills, D.M. & Zucker-Franklin, D. (1969) *Am. J. Path. 54*, 147-166.
5. Van Berkel, Th. J.C. (1974) *Biochem. Biophys. Res. Commun. 61*, 204-209.
6. Van Berkel, Th. J.C. & Kruijt, J.K.(1977) *Eur. J. Biochem. 73*, 223-229.
7. Van Berkel, Th. J.C. & van Van Tol, A. (1979) *Biochim. Biophys. Acta, in press.*
8. Van Berkel, Th.J.C., Kruijt, J.K., Slee, R.G. & Koster, J.F. (1977) *Arch. Biochem. Biophys. 179*, 1-7.

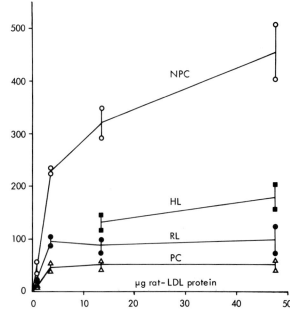

Fig. 3. Hydrolysis of rat iodinated LDL by homogenates of PC, NPC, total rat liver (RL) and human liver (HL), as a function of the amount of LDL apoprotein (μg/ml).

9. Van Berkel, Th.J.C., Koster, J.F. & Hülsmann, W.C. (1972) *Biochim. Biophys. Acta 276*, 425-429.
10. Jezyk, P.F. & Liberti, J.P. (1969) *Arch. Biochem. Biophys. 134*, 442-449.
11. Hommes, F.A., Draisma, M.I. & Molenaar, I. (1970) *Biochim. Biophys. Acta 222*, 361-371.
12. Sapag-Hagar, M., Marco, R. & Sols, A. (1969) *FEBS Lett. 3*, 68-70.
13. Crisp, D.M. & Pogson, C.I. (1972) *Biochem. J. 126*, 1009-1023.
14. Dunaway, G.A., Leung, G.L.Y., Cooper, M.D., Thrasker, J.R. & Wagle, S.R. (1978) *Biochem. Biophys. Res. Commun. 80*, 71-74.
15. Levin, M.J., Guillouzo, C., Hofmann, E. & Kahn, A. (1978), *to be published (cf. Abstr. 2941, 12th FEBS Meeting, Dresden)*.
16. Wincek, T.J., Hupka, A.L. & Sweat, F.W. (1975) *J. Biol. Chem. 250*, 8863-8873.
17. Wagle, S.R., Hofmann, F. & Decker, K. (1976) *Biochem. Biophys. Res. Commun. 71*, 857-863.
18. Van Berkel, Th.J.C. & Kruijt, J.K. (1977) in *Kupffer Cells and Other Liver Sinusoidal Cells* (Wisse, E. & Knook, D.L., eds.), Elsevier, Amsterdam, pp. 307-314.
19. Blouin, A., Bolender, R.P. & Weibel, E.R. (1977) *J. Cell Biol. 72*, 441-455.
20. Fukushima, M., Aihara, Y. & Ichiyama, A. (1978) *J. Biol. Chem. 253*, 1187-1194.
21. Daems, W.Th., Roos, D., Van Berkel, Th.J.C. & Van der Rhee, H.J. (1979) in *Lysosomes in Biology and Pathology*, Vol. 6 (Dingle, J.T. & Jacques, P.J., eds.), North-Holland, Amsterdam, *in press*.
22. Van Berkel, Th.J.C. & Kruijt, J.K. (1977) *Arch. Biochem. Biophys. 179*, 8-14.
23. Van Berkel, Th.J.C. & Koster, J.F. (1977) *as for* [18], pp. 299-306.
24. Cantrell, E. & Bresnick, E. (1972) *J. Cell Biol. 52*, 316-321.
25. Hupka, A.L. & Karler, R. (1973) *J. Ret. Soc. 14*, 225-241.
26. Gemsa, D., Woo, C.H., Fudenberg, H. & Schmid, R. (1974) *J. Clin. Invest. 53*, 647-651.
27. Bissell, D.M., Hammaker, L. & Schmid, R. (1972) *J. Cell Biol. 54*, 107-119.
28. Van Berkel, Th.J.C., Kruijt, J.K. & Koster, J.F. (1975) *Eur. J. Biochem. 58*, 145-152.
29. Knook, D.L. & Sleyster, E.Ch. (1976) *Mechanisms of Ageing and Development 5*, 389-397.
30. Berg, T. & Boman, D. (1973) *Biochim. Biophys. Acta 321*, 585-596.
31. Van Berkel, Th.J.C., Koster, J.F. & Hülsmann, W.C. (1977) *Biochim. Biophys. Acta 486*, 586-589.
32. Van Berkel, Th.J.C., Van Tol, A. & Koster, J.F. (1978) *Biochim. Biophys. Acta 529*, 138-146.
33. De Pijper, A.M. (1973) *Histochemie 37*, 197-206.

#B Separation Approaches, for Liver and other Cell Sources

#B-1
DISAGGREGATION AND SEPARATION OF RAT LIVER CELLS

PER O. SEGLEN
Department of Tissue Culture,
Norsk Hydro's Institute for Cancer Research,
The Norwegian Radium Hospital,
Montebello, Oslo 3, Norway.

The following procedure, entailing 2-stage disaggregation as amplified in the text and Figs. 5 and 6, is recommended wherever feasible. —

Source   *Liver from fasted rats.*

Dissocia-   *Perfusion (first with Ca^{2+}-free buffer, non-recircula-*
tion   *ting, then with collagenase + Ca^{2+}, recirculating.*
Final dissociation by combing.

Crude   *Intact hepatocytes (80%), damaged hepatocytes (6%),*
product   *Kupffer cells (8%), endothelial cells (6%).*

Separations *(1) Purification of hepatocytes by differential centri-*
(alterna-   *fugation. — Table 3. Purity 98-99%.*
tives) and *(2) Isopycnic separation on Metrizamide gradients*
products   *(analytical). Bands consisting of intact hepatocytes,*
damaged hepatocytes, and non-parenchymal cells
(heterogeneous).
(3) Preparative (isopycnic) isolation of non-parenchy-
mal cells on Metrizamide cushions. Yield, ∿30%,
purity ∿100%.
(4) Separation (isopycnic) of intact and damaged
hepatocytes on gradients of Metrizamide **and** *Percoll.*
Fractions 100% pure.
(5) Purification of anucleate hepatocytoplasts by
cytochalasin treatment, homogenization and isopycnic
separation on Metrizamide cushions.

Comments   *See CONCLUSIONS at end of text.*

The preparation of isolated cells from a tissue poses a problem with two major aspects (Fig. 1). The first concerns the disso-ciation (dispersion) of the tissue into a suspension of single

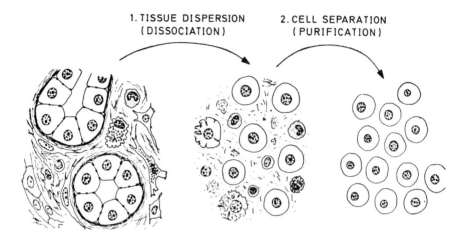

Fig. 1. The two major aspects of cell preparation.

cells and other tissue elements; the second concerns the purification of the initial cell suspension to leave intact cells of
the desired type, i.e. the removal of unwanted cell types,
damaged cells, subcellular debris and extracellular material.

Methods for tissue **dis**persion can generally be classified
into three categories: *mechanical, chemical* and *enzymatic*
methods. The various methods will now be briefly reviewed, and
the most successful one to date — the collagenase perfusion
technique for preparation of liver cells — will be described in
some detail. Among the numerous methods available for cell
separation, only separation according to cell density, i.e.
isopycnic banding   (e. g. in Metrizamide gradients), will be
dealt with here.   Other separation approaches are discussed
elsewhere in this volume.

CELLULAR JUNCTIONS

Tissue dispersion involves both dissolving the extracellular
matrix and breaking cell-to-cell contacts without breaking the
cell membrane. Cells in a tissue are held together by several
different types of junctions: desmosomes, tight junctions,
intermediate junctions and gap junctions (*see also* W.H. Evans,
this vol.). Fig. 2 shows several types of junctions between
adjacent hepatocytes in intact liver tissue. Each type of junction has its unique characteristics, and finding methods for the
specific cleavage of each one of them remains an essential problem in tissue dispersion technology. Desmosomes can be cleaved
through extraction of intercellular Ca^{2+} *(see below)*, but no

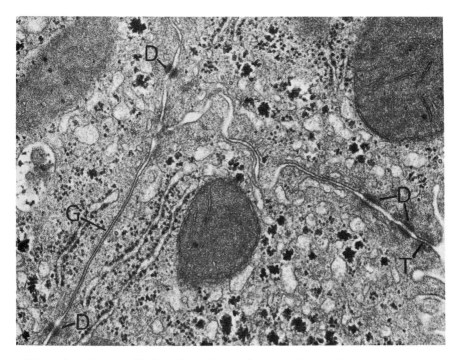

Fig. 2. Intercellular junctions in rat liver tissue.
On right, part of a bile capillary, sealed off by a junctional
complex consisting of a tight junction (T) and two desmosomes
(D); *on left,* a gap junction (G). x 50,000.

satisfactory methods are known for the cleavage of tight or gap
junctions. However, the cells are apparently capable of being
torn apart at these sites; the whole junction is left attached
to the surface of one cell, with re-sealing of the plasma membrane at the site of junction removal on the adjacent cell [1].

MECHANICAL DISPERSION METHODS

Table 1a lists a variety of mechanical methods which have been
used for the dissociation of different tissues. Several of
these can be very useful when used in conjunction with other
types of treatment, since some mechanical handling is always
necessary to disperse a tissue. However, it should be emphasized that a mechanical method alone is not sufficient to produce intact single cells. Tearing cells apart solely by mechanical force will result in the breakage of cell membranes
rather than in the cleavage of cell junctions.

Table 1. Non-enzymic methods for tissue dissociation.

Approach	Implementation [& ref.]
(a) *Mechanical methods*	
Mincing/slicing - - - -	scissors [2], razor blades [3, 4], tissue slicer [5, 6]
Homogenization - - - -	grinding, Waring blendor, pestle homogenizer [7, 8]
Scraping/stripping - - -	razor blade, rotating cylinder [9,10]
Sieving - - - - - -	tissue press [11], metal screen [2, 12, 13, & F. Ungar *et al., this vol.*]
Pipetting - - - - - -	pumping in syringe or pipette [4, 14; & F. Ungar *et al., this vol.*]
Shaking - - - - - -	e.g. with glass beads [3]
Teasing/combing - - - -	spatula [1], comb or forceps [15]
Vibration - - - - - -	ultrasonic probe [16] or other vibrating devices [17]
Microdissection - - - -	[14]
(b) *Chemical methods*	
Tetraphenylboron - - - -	K^+-chelator [18], probably ineffective [19]
Extreme pH - - - - -	acidic [4] or basic [20] medium; probably ineffective
Hyperosmolality - - - -	[20, 21]
Lectins, ligands - - -	competitive binding to junctional receptors; not sufficiently investigated
Divalent ion chelators -	citrate [7, 8], EDTA [7, 22], EGTA [23]

CHEMICAL DISPERSION METHODS

The only chemical agents (Table 1b) which have been convincingly proved to be efficacious in tissue dissociation are the divalent ion chelators (citrate, EDTA and the Ca^{2+}-specific chelator EGTA). One of their mechanisms of action has recently been elucidated, *viz.* the cleavage of desmosomes (Fig. 3). Liver perfusion experiments showed that the extraction of extracellular Ca^{2+} (with chelators or Ca^{2+}-free medium) promoted dispersion, in an irreversible manner, which was ascribed to the removal of a Ca^{2+}-dependent adhesion factor [19, 23]. Electron microscopic studies

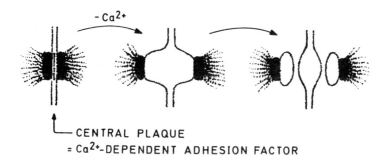

$-Ca^{2+}$

CENTRAL PLAQUE
$= Ca^{2+}$-DEPENDENT ADHESION FACTOR

Fig. 3. Desmosome cleavage upon removal of Ca^{2+} (by perfusion with Ca^{2+}-free buffer or chelators such as EGTA.
The Ca^{2+}-dependent central plaque material is extracted, where-upon the hemidesmosomes move apart and become internalized.

---

have demonstrated that the desmosomal 'central plaque', i.e. the material in the intercellular cleft between the two hemi-desmosomes, is extracted under such conditions [1, 22, 24, 25]. This results in cleavage of the desmosome and subsequent interiorization of the hemidesmosomes (Fig. 3), accounting for the irreversibility of the effect of Ca^{2+} removal.

The Ca^{2+}-requirement of liver desmosomes is ion-specific, i.e. Ca^{2+} cannot be replaced by Mg^{2+} [19]. However, the attachment of rat hepatocytes *in vitro* to a substratum of collagen or fibronectin is better supported by Mg^{2+} than by Ca^{2+} [26]. The possibility should therefore be considered that Mg^{2+}-dependent junctions between the cell and the extracellular matrix (or between cells) may exist *in vivo* and that divalent chelators may have an effect on such junctions.

Other types of chemical treatment have been attempted and, for example, hypertonic sucrose solutions have been reported to cleave gap junctions by causing the cells to shrink [21]. Tetra-phenylboron is a potassium chelator which has been claimed to facilitate the dissociation of mouse liver tissue [18], but it was completely ineffective in the perfused rat liver [19]. A potentially useful approach, provided that the biochemistry of cellular junctions is further elucidated, might be the use of specific lectins or sugar ligands to prevent the reciprocal binding between the surface receptors of adjoining cells.

ENZYMATIC DISPERSION METHODS

In addition to the breakage of cell-to-cell contacts, isolation of cells from a tissue requires dissolution of the extracellular matrix. This can be achieved by enzymatic treatment. A number of proteolytic and carbo-hydrate-degrading enzymes thought capable of degrading the extra-cellular proteoglycan matrix have been used, either singly or in various combinations (Table 2). Unspecific protease mixtures such as crude trypsin or pro-nase are effective in many cases, especially in loose, mesenchymal tissues of the type found in the embryo. In crude pancreatic protease, *trypsin* and *elastase* have been shown to be the active components, whereas other enzymes such as amylase and lipases are deleterious and may cause cell damage [28, 37, 40].

Unspecific proteases are less effective in adult tissues and may

Table 2. Enzymes used for tissue dissociation.

Enzyme	Amplification [& ref.]
Unspecific proteases	
Trypsin, crude	pancreatic enzyme mixture [27, 28]
Pancreatin	*ditto* [28]
Viokase	*ditto* [29, 30]
Pronase	bacterial protease mixture [31-33]
Trypsin, pure [34, 35]	
Chymotrypsin [29, 30]	
Chymopapain [32]	
Specific proteases	
Collagenase [1,13,15,24,29,32,35,36]	
Elastase [28,32,36,37]	
Other enzymes	
Hyaluronidase [1,24,29,36,38]	
Lysozyme [39]	
Neuraminidase [34]	
Deoxyribo-nuclease [13,32,34,35]	
Ribonuclease [35]	

inflict considerable damage upon epithelial cells. In fact, selective destruction of hepatocytes by pronase is the basis of certain methods for the purification of non-parenchymal cells from the liver [33, 41]. For adult epithelial tissues, *collage-nase* is the enzyme of choice. It will effectively dissolve the extracellular collagen/fibronectin network, and may have a 'softening' effect on certain cellular junctions [25], while leaving the cell surface unharmed as indicated by the persistence of intact surface receptors of various kinds [42-45]. Elastase may be particularly useful in elastin-rich tissues such as the lung [32].

The utility of other enzymes is more doubtful. Lysozyme, which has been reported to disperse liver tissue [39], has no effect in the perfused liver (Fig. 4). (Trypsin is likewise in-effective towards liver, while both crude and purified collage-nase disperse the tissue effectively, Fig. 4). *Deoxyribonuclease*

Fig. 4. Enzymatic dispersion of the isolated rat liver.
Perfusion of the liver, first with Ca²⁺-free buffer and then with Ca²⁺-activated crude (●) or purified collagenase (o), results in swelling of the tissue, whereas trypsin (Δ) or lysozyme (▲) has little or no effect. *From [46].*

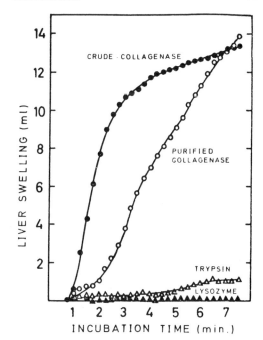

may be used to digest DNA released from damaged cells; this sticky material may otherwise cause cell aggregation.

COLLAGENASE PERFUSION

For both chelators and enzymes it **is** imperative that the dispersing agent gain access to all the cells in a tissue, and in the case of tissue slices or minces, poor penetration is usually the factor precluding a successful dispersion. Excellent penetration can be achieved, however, in those cases where it is possible to *perfuse* the tissue with the dissociating agents through the vasculature. With a continuous perfusion it may be possible to maintain optimal conditions for a sufficiently long time to disperse the tissue completely.

The best tissue dispersion yet achieved has been by the *collagenase perfusion* technique, first applied to the rat liver by Berry & Friend [1]. Subsequent developments of the method [15] have elucidated the principles and essential parameters involved, and under optimal conditions it is now possible to disperse the liver so effectively that the yield of hepatocytes approaches 100%, of which about 95% are intact [46]. By this procedure the isolated liver is perfused at 37° in two stages: in the first stage, extracellular Ca²⁺ is extracted (with Ca²⁺-free buffer or EGTA) in order to cleave the desmosomes, and in the second stage, the extracellular matrix is digested by collagenase (Ca²⁺ being put back to activate the enzyme). Since the method may be generally applicable to most tissues, a brief description of the procedure will be given. Further details can be found in refs. [15, 46].

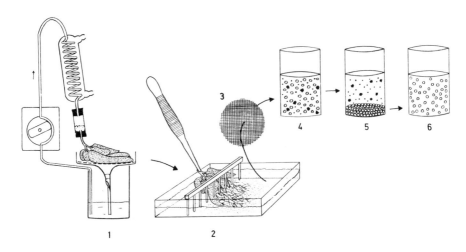

Fig. 5. Schematic outline of the collagenase perfusion technique.
The liver is dispersed by perfusion with Ca^{2+}-free buffer (10
min) and Ca^{2+}-activated collagenase (10 min) at 37° in a simple
perfusion apparatus (1), then transferred to a petri dish and
completely dissociated by means of combing (2). After pre-
incubation and filtration (3), intact hepatocytes are purified
by differential centrifugation (4-6).

Fig. 5 gives a schematic outline of the whole procedure for
isolation of hepatocytes by collagenase perfusion. The liver is
perfused by means of a very simple apparatus (represented above
by 1) consisting of a pump which will deliver perfusate through
2 × 4 mm silicone rubber tubing at a rate of 50 ml/min; a water-
jacketed glass coil maintaining the perfusate temperature at 37°;
a bubble trap (made of a piece of wide silicone rubber tubing
with a stopper in each end) with a cotton-wool filter; a glass
dish with a conical outlet and a stainless-steel net to support
the liver; and a beaker serving as perfusate reservoir. A
piece of split silicone-rubber tubing attached to the dish outlet serves to aid the splash-free return of the perfusate to the
reservoir. Oxygenation of the perfusate has been found to be
unnecessary during the short time period (2 × 10 min) required
for liver dispersion [46].

The liver is cannulated (with a piece of nylon tubing) through
the *vena porta* under ether anaesthesia, and the first stage of
perfusion — i.e. with Ca^{2+}-free buffer (8.3 g NaCl, 0.5 g KCl,
2.4 g HEPES, 5.5 ml 1 M NaOH and H_2O *ad* 1000 ml; pH 7.4 at 37°) —
is immediately started while the liver is lying *in situ*. Buffer
(500 ml) is pumped through the liver in 10 min without re-circulation, and during this time period the liver is carefully excised

and placed in the dish. With this very efficient perfusion the
inclusion of chelators has been found to be unneccesary; however, when the perfusate volume is small and re-circulation is
used [23], or if ineffective Ca^{2+} extraction is to be expected,
e.g. during the incubation of tissue slices, a Ca^{2+} chelator
such as EGTA is recommended.

In the second stage, the reservoir is filled with 50 ml of
collagenase buffer (3.9 g NaCl, 0.5 g KCl, 0.7 g $CaCl_2 \cdot 2 H_2O$,
24.0 g HEPES, 66 ml 1 M NaOH, 0.5 g collagenase Sigma type I,
and H_2O *ad* 1000 ml; pH 7.6 at 37°), which is re-circulated for
10 min at 50 ml/min. The re-addition of Ca^{2+} is necessary
because collagenase is a Ca^{2+}-requiring enzyme, and it is per-
missible because the desmosomes at this stage should be irrever-
sibly cleaved. If they are not, e.g. due to a too brief or
otherwise imperfect perfusion, Ca^{2+} addition will stop further
cleavage. For this reason some investigators prefer not to re-
add Ca^{2+}, relying on the ability of collagenase to derive
necessary Ca^{2+} from the tissue (thereby possibly also aiding
desmosome cleavage). However, it should be stressed that the
best and most consistent results are obtained if one treats
these two stages — junction cleavage and matrix degradation —
as distinct, and ensures that optimal conditions prevail at
each stage.

After the perfusion with collagenase, during which the liver
swells visibly to about twice its original size, it is transferred to a square petri dish where the liver capsule is disrupted and the cells disentangled from the vascular and connective tissue network by gentle combing with a stainless-steel dog
comb (2, Fig.5).The resulting cell suspension is usually pre-
incubated at 37°for 30 min in a large (20 cm) shaking petri
dish; although this step is not absolutely necessary, it will
cause the intact cells to round up and the damaged cells to
become lighter, both of which will facilitate the subsequent
purification. After pre-incubation the cells are filtered
through 250 µm and 100 µm nylon mesh to remove coarse tissue
fragments and aggregated cells (3), and then centrifuged 4 times
in flat-bottomed beakers (0-4°) at a centrifugal force (20 *g*
for 2 min) which will selectively sediment the large, heavy intact hepatocytes (4-6).

CHARACTERIZATION OF THE CELL PREPARATION

A cell preparation can be adequately characterized in terms of
yield (relative to the initial amount of tissue), *purity* (i.e.
the relative contamination by unwanted cell types) and *viability*
(i.e. the relative number of *intact* cells of the desired type).
The ability of cells to exclude the dye trypan blue allows a
simple and rapid viability test (testing the structural integrity

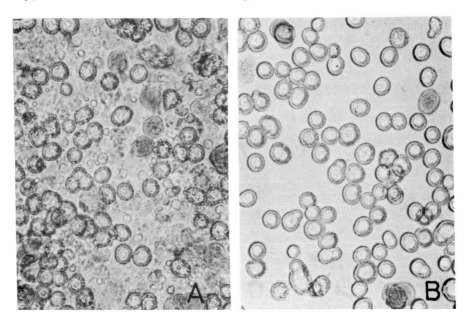

Fig. 6. Purification of hepatocytes by differential centrifugation.
The pre-incubated cell suspension is centrifuged 4 times at 40 *g*-
min in the cold. The supernatants are discarded, and the final
pellet is re-suspended and used for experiments.
A, first supernatant containing many small, non-parenchymal cells
and debris; B, final cell suspension. x 260.

of the plasma membrane) provided that it is stringently applied
[15]; other functional tests are discussed by H.A. Krebs *et al.*
(this vol.).

Since a final cell preparation is usually the result of both
tissue dispersion and cell purification procedures, the dispersion method should be evaluated on the basis of the *initial* cell
suspension obtained, i.e. before purification. The suspension
obtained after collagenase perfusion consists mostly of hepatocytes, 90-95% of which are intact, some 10-30% of small, non-
parenchymal cells (Kupffer cells and endothelial cells in almost
equal numbers [41]), capsule fragments and subcellular debris
(Fig. 6A). Disregarding the vascular and connective tissue rem-
nant (which constitutes about 15% of the liver weight), the
initial suspension represents a yield of virtually 100%.

Purification by the differential centrifugation procedure outlined above results in the removal of most of the debris and non-
parenchymal cells, and a pure preparation of, mainly, viable

Table 3. Purification of hepatocytes by differential centrifugation (differential pelleting).
The cells in the final suspension were sedimented 4 times at the
$g \cdot min$ value indicated. The cells in the initial cell suspension ($\sim$100% of the hepatocytes) represented 86% of the liver
weight. *Modified from ref. [46]. EDITOR'S NOTE. — The term
'parenchymal' as used now and by some other authors (e.g. T.J.C.
van Berkel, this vol.) excludes sinusoidal cells, but could
reasonably include them as suggested in the Preface.*

	Initial cell suspension	Final cell suspension	
		25 $g \cdot min$	90 $g \cdot min$
Parenchymal cell (hepa-tocyte) yield, % *of initial*	$\sim$100	42	64
Viability, % *intact parenchymal cells*	93	93	95
Non-parenchymal cells, *% of total cell number*	14	1.4	5.1

intact hepatocytes is obtained (Fig. 6B). Table 3 gives the
quantitative aspects of purification. With an initial viability
around 95% no further improvement of this parameter can be expected, but with lower initial viabilities there may be a considerable enrichment of intact cells in the final preparation. The
final yield of hepatocytes ($\sim$50% by standard procedures, i.e.
40 $g \cdot min$) can be increased by increasing the sedimentation force,
but this results in a greater contamination by non-parenchymal
cells. The usual contamination of 2% by cell number in the
standard preparation represents only about 0.1% by cell mass,
due to the small size of the non-parenchymal cells [41].

METHODS FOR CELL PREPARATION

Numerous methods have been developed for the separation and
purification of different cell types (Table 4). Differential
centrifugation — i.e. separation according to cell size — has
already been briefly discussed. Other methods are dealt with
elsewhere in this volume. In our laboratory we have concentrated on methods for separation according to cell density, i.e.
isopycnic banding techniques using primarily the gradient medium
Metrizamide [53, 55].

DENSITY GRADIENT MATERIALS FOR CELL SEPARATION

For a gradient substance to be useful for cell separation, it
must be capable of forming solutions of a density greater than

that of the cells, without exceeding the physiological osmolality
of about 300 m-osmol/kg (which would cause cell shrinkage).
Table 5 compares the properties of three such gradient substances: Metrizamide (Nyegaard & Co., Oslo, Norway), Percoll and
Ficoll (Pharmacia, Uppsala, Sweden). *Metrizamide* is a small-
molecular weight iodinated benzamido-glucose derivative [64],
Percoll is a colloidal suspension of silica particles coated
with polyvinylpyrrolidone (H. Pertoft *et al., this vol.*), and
Ficoll is a high-molecular weight sucrose polymer. The substances are compared at a solution density of 1.13 g/ml, which
is the maximum density attainable with commercially available
Percoll; with Metrizamide isotonic solutions of densities up to
1.21 g/ml can be prepared [55]. At 1.13 g/ml, which is just
sufficient to float hepatocytes (density 1.12) [53], Ficoll is
extremely viscous, and has also greatly exceeded the physiological osmolality of 300 m-osmol/kg; it is therefore useless for
liver-cell separation. Metrizamide, on the other hand, has a
very low viscosity, and Percoll has a very low osmolality. For
the preparation of gradients, a working solution of buffered 30%
(w/v) Metrizamide has been used (300 g Metrizamide, 0.5 g KCl,
0.18 g $CaCl_2 \cdot 2 H_2O$, 2.4 g HEPES, 5.5 ml M NaOH and H_2O *ad*
1000 ml; pH 7.4 at 37° and 260 m-osmol/kg), diluted as required
with a corresponding buffer (7.14 g NaCl, 0.5 g KCl, 0.18 g
$CaCl_2 \cdot 2 H_2O$, 2.4 g HEPES, 5.5 ml 1 M NaOH and H_2O *ad* 1000 ml;
pH 7.4 at 37° and 260 m-osmol/kg). The same composition of
buffer and salts was added to the commercially available Percoll
solution.

ISOPYCNIC SEPARATION OF LIVER CELLS

Near-linear Metrizamide gradients were generated by the method
of Stone [65], i.e. by layering 3 or 4 Metrizamide solutions of
different densities on top of each other and allowing diffusion
between the layers to occur for 90 min at 4° with the stoppered
tubes in a horizontal position [53]. In most experiments cells
were subsequently layered on top of the gradient (the cell sus-
pension/Metrizamide interface being rendered smooth by gentle
mixing with a spiral device) and the tubes were centrifuged at
0-4° for 10 min at 3000 rev/min. In one experiment, however
(Fig. 7), the initial (heterogenous) cell suspension was pre-
mixed into the gradient layers in order to permit the loading of
a large cell quantity while avoiding interface packing; in this
way an adequate number of non-parenchymal cells was present and
the counted cell numbers were significant. In such a gradient
the hepatocytes formed a well-defined, homogeneous peak at a
density of 1.12 g/ml, while the non-parenchymal cells distributed more heterogeneously, with a major peak at 1.07 (Fig. 7).

Table 4 *(opposite)*. Methods for separation and purification of cells.
ADDED BY EDITOR.—Ligand methods, e.g. [59]; *ion-exchangers,* #D-**2**;
magnetic methods, pp. 67, 232.

Approach	Amplification [& ref.]
Separation according to cell size	
Filtration - - - - - -	mesh openings of successively diminishing size
Differential centrifugation (differential pelleting)	multiple sedimentations, 'washing' [46, and R.J. Hay, *this vol.*]
Rate (zonal) centrifugation	shallow density gradients used for stabilization; no equilibrium reached [47, 48]
Velocity sedimentation - -	sedimentation at unit *g* in shallow stabilizing gradients [49, 50, and W.S. Bont & J.E. de Vries, *this vol.*]
Elutriation - - - - - -	sedimentation-rate equilibrium in counterflow velocity gradients [51, 52, and D.L. Knook & E.Ch. Sleyster, *this vol.*]
Separation according to cell density	
Isopycnic banding - - - -	centrifugation to density equilibrium in density gradients [13, 53, and H. Pertoft *et al.*, *this vol.*]
Dense cushion - - - - -	batch procedure for separation at one specified density [54, 55]
Flotation - - - - - - -	batch procedure for separation at one specified density [55, and J.P. Luzio, *this vol.*]
Separation according to cell surface properties	
Phase distribution - - - -	single-step or counter-current [56, and I.A. Sutherland, *this vol.*; E. Eriksson & G. Johansson, *this vol.*]
Electrophoresis - - - - -	[57, and H.-G. Heidrich & K. Hannig, *this vol.*]
Column fractionation - - -	beads or fibres [58, 59]
Selective attachment - - -	in tissue culture [41, 60]
Separation according to other properties	
Electronic sorting - - - -	e.g. flow-cytofluorimetric [61]
Selective destruction - - -	e.g. enzymatic [33, 41]
Selective growth or survival	Auxotrophy [62]; toxin resistance [63]

Table 5. Physicochemical properties of gradient media.

Solute	Concentration, % (w/v)	Density, g/ml	Viscosity, cps	Osmolality, m-osmol/kg
Metrizamide	25	1.13	2	210
Percoll	26	1.13	10	15
Ficoll	42	1.13	>100	500

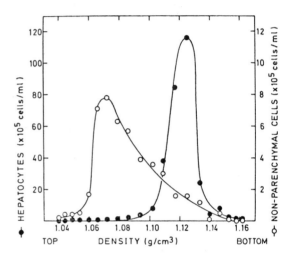

Fig. 7. Distribution of rat liver cells in a Metrizamide density gradient. An unpurified cell suspension (primary dissociate) was incorporated into a diffusion-generated 5-30% Metrizamide gradient, and centrifuged to equilibrium (20 min at 5000 rev/min) at 4°. Hepatocytes. ●; non-parenchymal cells, o . *Note 10-fold scale difference.*

In contrast to this result, other authors have reported a density heterogeneity among rat hepatocytes, with several resolvable peaks (bands) in density gradients [25, 66, 67]. As shown in Fig. 8, increased resolution of the Metrizamide gradient still resulted in only one homogeneous band. (This and subsequent experiments were done with hepatocytes purified by differential centrifugation, which always exhibited the same density distribution as the hepatocytes in the unpurified, initial cell suspension.)

Hepatocytes prepared from male or female rats, fed or fasted rats, young or old rats have likewise been found to give only one band in Metrizamide density gradients. In cell preparations from young rats a faint second band of light cells may sometimes be seen (Fig. 9A); however, microscopic examination reveals that this band consists of Kupffer cell aggregates of approximately the size of single hepatocytes (Fig. 9B). Although

Fig. 8. Banding of hepatocytes
in Metrizamide gradients.
A, 6-26% Metrizamide; B, 10-22.5%
Metrizamide. *The sharp, thin band
near the top of the tubes represents
light reflection in the meniscus.*

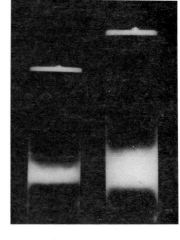

Fig. 9 *(bottom, right)*. Banding
of hepatocytes and contaminating
Kupffer cell aggregates in a
Metrizamide gradient.
A, a 6-26% Metrizamide gradient
containing cells purified from a
young-adult (220 g), well-fed
female rat. *Apart from the cell
bands, some unspecific light reflec-
tions appear at the meniscus and at
the tube bottom.*
B, cells recovered
from the upper
faint band, i.e.
Kupffer-cell
aggregates; C,
cells recovered
from the lower,
major band, i.e.
pure hepatocytes.
x 260.

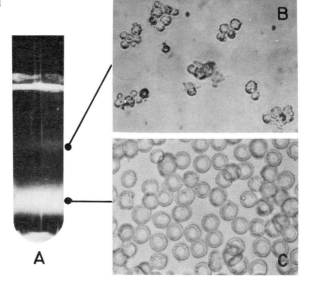

the heterogeneity reported by other workers remains unexplained,
our data strongly indicate that rat hepatocytes form a single,
uniform population with respect to cell density.

PURIFICATION OF NON-PARENCHYMAL LIVER CELLS

It is apparent from Fig. 7 that the density difference between
hepatocytes and non-parenchymal liver cells is sufficient to
allow a preparative purification of non-parenchymal cells with

Fig. 10. Purification
of non-parenchymal
cells by centrifuga-
tion above a dense
Metrizamide cushion.
A 2 ml portion of un-
purified cell suspen-
sion (first super-
natant) was layered
on top of 5 ml of
15% Metrizamide
(density: 1.08 g/ml),
mixed at the inter-
face, and centrifuged
for 5 min at 5000 rev/
min (A). *Hepatocytes*

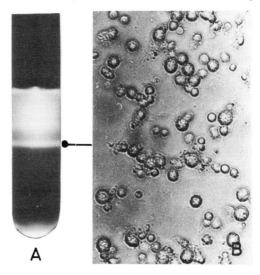

sediment to the
bottom of the tube **A**
and the debris re-
mains in the top
layer (causing considerable light reflection), while non-
parenchymal cells form a band at the interface.   . B, Non-
parenchymal cells recovered as a pure preparation from the
interface.   x 500.

---

good yield. If a cell suspension is layered above a Metrizamide layer with a density of 1.08 g/ml, all the hepatocytes
penetrate the layer, while most of the non-parenchymal cells
are retained at the interface [55]. As shown in Fig. 10, the
preparation of non-parenchymal cells recovered from such an
interface is 100% pure. This provides a very simple method for
the purification of non-parenchymal cells, obviating the need
to use surface receptor-destructive agents such as pronase [33,
41].

SEPARATION OF INTACT FROM DEAD CELLS

In 5-30% (w/v) Metrizamide gradients such as the one depicted
in Fig. 7, damaged cells were found to sediment to the bottom
of the tube [53], and a single-step method (centrifugation of
cells onto a cushion of Metrizamide of an appropriate density)
for the general separation of intact from damaged cells was subsequently suggested [15, 55]. Apparently Metrizamide is excluded from cells with an intact plasma membrane while it penetrates into damaged cells; the contribution made by cytosol
water to the overall cellular density of the latter is thus
eliminated. In Percoll, on the other hand, damaged cells were
found, surprisingly, to have a *lower* density than intact cells
(H. Pertoft *et al.*, *this vol.*).

Fig. 11. Separation of
intact and damaged cells
on gradients of Metriza-
mide and Percoll.
Complete cell damage
(100% staining with trypan
blue) was inflicted by
subjecting a cell sample
to a single freeze/thaw
cycle in liquid N_2, and
intact and damaged cells
alone or in a 50 : 50 mixture were centrifuged to
equilibrium (5 min; 500
rev/min) in gradients of
Metrizamide (3→40%, w/v)
or Percoll (10→75%, v/v,
placed on top of a
30-40% Metrizamide gradient).

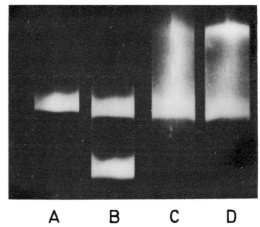

A   B   C   D

A, Intact hepatocytes in a Metrizamide gradient; B, a 50 : 50
mixture of intact and damaged hepatocytes in a Metrizamide gradient; C, intact hepatocytes in a Percoll gradient; D, a 50 : 50
mixture of intact and damaged cells in a Percoll gradient.
Percoll gradients appear opaque due to light reflection.

Fig. 11 compares this striking difference in the separation
properties of the two media, using gradients extending to a
density of 1.21 g/ml (40% Metrizamide). In order to facilitate
comparison, the Percoll gradients (1.00-1.13 ml) have been generated on top of Metrizamide gradients (1.13-1.21). Intact
hepatocytes can be seen to form a single band both in Metrizamide (A) and Percoll (C) gradients at the usual density of 1.12
g/ml. A 50 : 50 mixture of damaged (frozen/thawed) and intact cells
gave two well-separated bands in both types of gradient, but in
Metrizamide (B) the damaged cells were much heavier (∿1.17 g/ml)
than the intact cells, while in Percoll (D) they were much
lighter (∿1.05). Photographs of cells from the two Metrizamide
gradient bands show that they contain 100% intact cells (Fig.
12A) and 100% damaged cells (Fig. 12B), respectively, i.e. a
remarkably clear-cut separation. The pictures also reveal that
the damaged cells are swollen (Fig. 12B), and thus have a larger
water space than the intact cells. If the silica particles in
the Percoll gradient are too large to penetrate the structure of
damaged cells, these cells will acquire a lower density than intact cells (and Metrizamide-filled damaged cells) simply because
they contain more water.

The separation of intact cells from damaged cells by centrifugation above a dense Metrizamide cushion has been found in our

Institute to be an extremely useful general method for the purification of cells from preparations which initially contain many
damaged cells, e.g. trypsinized human biopsy specimens. Most
cells have a lower density than hepatocytes, and a layer of 20%
Metrizamide (1.105 g/ml; 40% aqueous stock solution, 1.21 g/ml =
isotonic, diluted with an equal amount of any isotonic buffer of
medium) will generally be sufficient for a quantitative recovery
of intact cells from the interface. A precaution which should
always be taken is to smoothen the interface (by gentle swirling,
or stirring with a spiral device) before centrifugation; otherwise the cells may be crushed against the interface or form a
compressed layer through which the damaged cells do not penetrate. If the preparation contains much subcellular debris,
the supernatant above the layer of intact cells may be removed,
and the tubes centrifuged a second time at high speed, to drive

Fig. 12. Puri-
fied fractions
of intact and
damaged cells
from a Metriza-
mide gradient.
Cells from the
upper (A) and
lower (B) band
in the Metriza-
mide gradient
shown in Fig.
11B were re-
covered and
stained with
trypan blue.
*The upper band
contains 100%
intact cells;
the lower band
100% damaged
cells.* x 260.

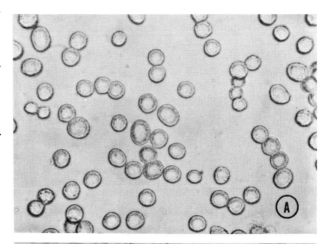
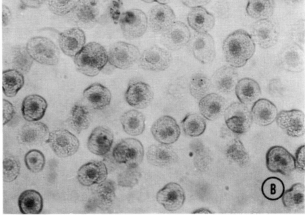

the debris (which is heavier than intact cells) into the Metrizamide layer. Such a procedure has been successfully used to prepare pure, intact anucleate cytoplasts from cytochalasin-treated, homogenized hepatocytes [55].

CONCLUSIONS

Isopycnic banding, especially in Metrizamide gradients, is a useful approach for the preparative and analytical separation of different cell types, and for the removal of damaged cells from a cell preparation. For analytical work, it is extremely important to use *isotonic* gradients (of a viscosity which permits the cells to reach equilibrium), and to avoid interface artefacts. For the analysis of cell populations comprising tissues it is also imperative to use a dispersion procedure which will produce non-aggregated, intact single cells in a nearly quantitative yield. Whenever feasible, the two-stage collagenase perfusion technique is recommended, or at least application of the principles of this method: first desmosome cleavage by the removal of Ca^{2+}, then digestion of the extracellular matrix with Ca^{2+}-activated collagenase.

Acknowledgements

I thank Amy Munthe-Kaas and Paul Gordon for their experimental and technical collaboration, and Barbara Schüler for providing the electron micrograph. The work was supported by a Norwegian Cancer Society grant.

References

1. Berry, M.N. & Friend, D.S. (1969) *J. Cell Biol. 59*, 722-734.
2. Schrek, R. (1944) *Arch. Pathol. 37*, 319-327.
3. St. Aubin, P.M.G. & Bucher, N.L.R. (1952) *Anat. rec. 112*, 797-809.
4. Longmuir, L.S. & ap Rees, W. (1956) *Nature (Lond.) 177*, 997.
5. Zaroff, L., Sato, G. & Mills, S.E. (1961) *Exp. Cell Res. 23*, 565-575.
6. Howard, R.B. & Pesch, L.A. (1968) *J. Biol. Chem. 243*, 3105-3109.
7. Anderson, N.G. (1953) *Science (Wash.) 117*, 627-628.
8. Jacob, S.T. & Bhargava, P.M. (1962) *Exp. Cell Res. 27*, 453-467.
9. Hülsmann, W.C., van den Berg, J.W.O. & de Jonge, H.R. (1974) *Meth. Enzymol. 32*, 665-673.
10. Sjöstrand, F.S. (1968) *J. Ultrastruct. Res. 22*, 424-442.
11. Tsai, C.M., Best, N. & Ebner, K.E. (1966) *Exp. Cell Res. 44*, 332-340.
12. Kaltenbach, J.P. (1954) *Exp. Cell Res. 7*, 568-571.
13. Fong, J.S.C. & Drummond, K.N. (1974) *Meths. Enzymol. 32*, 653-658.

14. Chen, J.S. & Levi-Montalcini, R. (1970) *Proc. Natl. Acad. Sci. U.S.A. 66,* 32-39.
15. Seglen, P.O. (1976) in *Methods in Cell Biology,* Vol. 13 (Prescott, D.M., ed.), Academic Press, New York, pp. 29-83.
16. Sanford, W.C. (1974) *In Vitro 10,* 281-283.
17. Parry, J.S., Cleary, B.K., Williams, A.R. & Evans, D.M.D. (1971) *Acta Cytol. 15,* 163-166.
18. Rappaport, C. & Howze, G.B. (1966) *Proc. Soc. Exp. Biol. Med. 121,* 1010-1016.
19. Seglen, P.O. (1973) *Exp. Cell Res. 76,* 25-30.
20. McLimans, W.F. (1969) in *Axenic Mammalian Cell Reactions* (Tritsch, G.L., ed.), Marcel Dekker, New York, pp. 307-367.
21. Goodenough, D.A. & Gilula, N.B. (1974) *J. Cell Biol. 61,* 575-590.
22. Coman, D.R. (1954) *Cancer Res. 14,* 519-521.
23. Seglen, P.O. (1972) *Exp. Cell Res. 74,* 450-454.
24. Amsterdam, A. & Jamieson, J.D. (1974) *J. Cell Biol. 63,* 1037-1056.
25. Drochmans, P., Wanson, J.C. & Mosselmans, R. (1975) *J. Cell Biol. 66,* 1-22.
26. Seglen, P.O. & Fosså, J. (1978) *Exp. Cell Res. 116,* 199-206.
27. Moscona, A., Trowell, O.A. & Willmer, E.N. (1965) in *Cells and Tissues in Culture,* Vol. 1 (Willmer, E.N., ed.), Academic Press, London and New York, pp. 49-60.
28. Speicher, D.W. & McCarl, R.L. (1974) *In Vitro 10,* 30-41.
29. Vale, W., Grant, G., Amoss, M., Blackwell, R. & Guillemin, R. (1972) *Endocrinology 91,* 562-572.
30. Harary, I., Hoover, F. & Farley, B. (1974) *Meth. Enzymol. 32,* 740-745.
31. Wiebelhaus, V.D., Blum, A.L. & Sachs, G. (1974) *Meth. Enzymol. 32,* 707-717.
32. Gould, K.G., Clements, J.A., Jones, A.L. & Felts, J.M. (1972) *Science 178,* 1209-1210.
33. Roser, B. (1968) *J. Reticuloendoth. Soc. 5,* 455-471.
34. Hopkins, C.B. & Farquhar, M.G. (1973) *J. Cell Biol. 59,* 276-308.
35. Barofsky, A.-L., Feinstein, M. & Halkerston, I.D.K. (1973) *Exp. Cell Res. 79,* 263-274.
36. Kraehenbuhl, J.P. (1977) *J. Cell Biol. 72,* 390-405.
37. Phillips, H.J. (1972) *In Vitro 8,* 101-105.
38. Kimmich, G.A. (1970) *Biochemistry 9,* 3659-3668.
39. Hommes, F.A., Draisma, M.I. & Molenaar, I. (1970) *Biochim. Biophys. Acta 222,* 361-371.
40. Speicher, D.W. & McCarl, R.L. (1978) *In Vitro 14,* 849-853.
41. Munthe-Kaas, A.C., Berg, T., Seglen, P.O. & Seljelid, R. (1975) *J. Exptl. Biol. Med. 141,* 1-10.
42. Le Cam, A., Guillouzo, A. & Freychet, P. (1976) *Exp. Cell Res. 98,* 382-395.
43. Fouchereau-Peron, M., Rançon, F., Freychet, P. & Rosselin, G. (1976) *Endocrinology 98,* 755-760.

44. Postel-Vinay, M.-C. & Desbuquois, B. (1977) *Endocrinology 100*, 209-215.
45. Ranke, M.B., Stanley, C.A., Tenore, A., Rodbard, D., Bongiovanni, A.M. & Parks, J.S. (1976) *Endocrinology 99*, 1033-1045.
46. Seglen, P.O. (1973) *Exp. Cell Res. 82*, 391-398.
47. Pretlow, T.G. & Williams, E.E. (1973) *Anal. Biochem. 55*, 114-122.
48. Warmsley, A.M.H. & Pasternak, C.A. (1970) *Biochem. J. 119*, 493-499.
49. Miller, R.G. (1973) in *New Techniques in Biophysics and Cell Biology* (Pain, R.H. & Smith, B.J., eds.), Wiley, London, pp. 87-112.
50. Zeiller, K., Hansen, E., Leihener, D., Pascher, G. & Wirth, H. (1976) *Hoppe-Seyl. Z. Physiol. Chem. 357*, 1309-1319.
51. Smith, D.F. (1969) *Exptl. Cell Res. 57*, 251-256.
52. Glick, D., von Redlich, D., Juhos, E.T. & McEwen, C.R. (1971) *Exptl. Cell Res. 65*, 23-26.
53. Munthe-Kaas, A.C. & Seglen, P.O. (1974) *FEBS Lett. 43*, 252-256.
54. Böyum, A. (1968) *Scand. J. Clin. Lab. Invest. 21, Suppl. 97*, pp. 1-109.
55. Seglen, P.O. (1976) in *Biological Separations in Iodinated Density-Gradient Media* (Rickwood, D., ed.), Information Retrieval, London, pp. 107-121.
56. Albertsson, P.-Å. (1969) in *Modern Separation Methods of Macromolecules and Particles* (Gerritsen, T., ed.), Wiley-Interscience, New York, pp. 105-120.
57. Zeiller, K., Löser, R., Pascher, G. & Hannig, K. (1975) *Hoppe-Seyl. Z. Physiol. Chem. 356*, 1225-1244.
58. Shortman, K., Williams, N., Jackson, H., Russell, P., Byrt, P. & Diener, E. (1971) *J. Cell Biol. 48*, 566-579.
59. Edelman, G.M., Rutishauser, U. & Millette, C.F. (1971) *Proc. Natl. Acad. Sci. U.S.A. 68*, 2153-2157.
60. Polinger, I.S. (1970) *Exp. Cell Res. 63*, 78-82.
61. Julius, M.H., Sweet, R.G., Fathman, C.G. & Herzenberg, L.A. (1975) in *Mammalian Cells: Probes and Problems* (Richmond, C.R., Petersen, D.F., Mullaney, P.F. & Anderson, E.C., eds.) Technical Information Center, Office of Public Affairs, U.S. Energy Research and Development Administration, Los Alamos, pp. 107-121.
62. Leffert, H.L. & Paul, D. (1972) *J. Cell Biol. 52*, 559-568.
63. Judah, D.J., Legg, R.F. & Neal, G.E. (1977) *Nature (Lond.) 265*, 343-345.
64. Hinton, R.H. & Mullock, B.M. (1976) in *Biological Separations in Iodinated Density-Gradient Media* (Rickwood, D., ed.), Information Retrieval, London, pp. 1-14.
65. Stone, A.B. (1974) *Biochem. J. 137*, 117-118.
66. Weigand, K., Otto, I. & Schopf, R. (1974) *Acta Hepato-Gastroenterol. 21*, 245-253.

67. Gumucio, J.J., De Mason, L.J., Miller, D.L., Krezoski, S.O. & Keener, M. (1978) *Am. J. Physiol. 234,* C102-C109.

#B-2

LIVER SINUSOIDAL CELLS: ISOLATION AND PURIFICATION BY CENTRIFUGAL ELUTRIATION

D. L. KNOOK and ELIZABETH Ch. SLEYSTER
Institute for Experimental Gerontology TNO,
Rijswijk, The Netherlands.

Source & types of cell	*Pre-perfused liver from 3-month (140-170 g) female BN/BiRij rats, or 2-month (20-24 g) male GRS mice: sinusoidal cells; purified Kupffer and endothelial cells.*
Liver perfusion	*Perfusion* in situ *with Gey's balanced salt solution (GBS) pH 7.4, 275 mOsm (for mouse 308 mOsm), 2 min, flow rate 10 (mouse, 5) ml/min; then with GBS + 0.2% pronase E (Merck), 1 min, 10 (mouse, 5) ml/min.*
Dissociation	*Small pieces of liver in GBS + 0.2% pronase, 37°: stir 60 min, keep at pH 7.4, filter through gauze, centrifuge at 300 g for 5 min to collect 'non-parenchymal' cells*.*
Separation	*Erythrocyte removal by centrifugation at 1400 g for 10 min through Metrizamide cushion, density 1.089 (mouse, 1.086) g/ml at 21°. Then centrifugal elutriation† (Beckman JE-6 rotor), 2550 rev/min: lymphocyte fraction obtained at flow rate (GBS) of 11.3 ml/min, endothelial cell fraction at 20.7 ml/min, Kupffer cell fraction at 38 ml/min.*
Product	*Rat or (parenthetically) mouse: 32.2 ±1.2 [S.E.] (15.3 ±1.5) × 10^6 non-parenchymal cells/g liver [99.6 ±4.2 × 10^6 cells/100 g rat]; 16.2 ±0.8 × 10^6 endothelial cells/g liver, purity 96%, viability*

---

*EDITOR'S NOTE (cf. Preface).— The term 'parenchymal' as used by some workers including E.R. Weibel encompasses not only hepatocytes but also sinusoidal cells; in the present paper the latter are included in the 'non-parenchymal' category.

†Cf. the account by I. Schulz and co-authors, *this vol.*

96%; 4.2 ±1.2 (4.0 ±0.9) × 10⁶ Kupffer cells/g
liver, purity 90% (92%), viability 93% (90%).

Comments &   *Yields of non-parenchymal cells and of Kupffer cells*
alternatives   *lower if collagenase used instead of pronase. For*
               endothelial cells, a passable alternative to centri-
               fugal elutriation is Metrizamide density-gradient
               centrifugation after in vivo *loading of the Kupffer*
               cells.

There is a fast growing interest in liver cells other than parenchymal cells [in the sense of hepatocytes — *Ed.*]. 'Non-parenchymal' (non-hepatocyte) cells are mostly sinusoidal cells, consisting of endothelial cells, Kupffer cells, fat-storing cells
and pit cells. Several biochemical and functional characteristics
of isolated sinusoidal cells have already been studied (T.J.C. van
Berkel, *this vol.*). Further studies on the characteristics of
the various sinusoidal cell types recently became feasible through
the development of techniques for preparing highly purified
Kupffer and endothelial cell suspensions from mouse and rat liver.
The procedure entails (1) the preparation of a non-parenchymal
cell suspension (mainly sinusoidal cells), and (2) further purification to give fractions consisting almost solely of Kupffer
cells or endothelial cells. The use for (2) of a method based on
centrifugal elutriation furnishes both cell types in a functionally and structurally intact state.

IDENTIFICATION OF KUPFFER AND ENDOTHELIAL CELLS

Identification of isolated Kupffer cells in non-hepatocyte cell
suspensions can be based on (a) a positive cytoplasmic staining
for peroxidatic activity, (b) ultrastructural characterization,
or (c) unique ability to phagocytose well defined and recognizable
large particles. These methods have recently been reviewed [1].
Endothelial cells can be directly identified only by (b). No
specific enzyme cytochemical marker is known for this cell type
[2], but the number of these cells in non-hepatocyte suspension
can be estimated indirectly by subtracting the number of cells
with positive peroxidatic activity by light microscopy, i.e.
Kupffer cells, from the number with a positive esterase activity,
i.e. Kupffer cells plus endothelial cells [2].

PREPARATION OF 'NON-PARENCHYMAL' (NON-HEPATOCYTE) CELLS

The several methods that have been described have recently been
compared [1, 3]. The methods most widely used today are based on
the selective digestion of parenchymal cells (hepatocytes) by
pronase (an enzyme combination), resulting in a suspension nearly
free of hepatocytes. Experimental details as outlined above have
been described elsewhere [3]. This method was originally described by Roser [4] for mouse liver and by Mills and Zucker-

Franklin [5] for rat liver. For some specialized studies on membrane characteristics collagenase may be preferred to pronase [1], although (with rat liver) it gives a worse yield and relatively fewer Kupffer cells [1, 3].

SELECTIVE ATTACHMENT OF KUPFFER CELLS

Assuming that Kupffer cells are the only cells in a non-hepatocyte cell suspension that attach during maintenance culture, they can thereby be separated out [2, 6]. The estimated yield of Kupffer cells is under 6×10^6 cells per g rat liver [1, 6]. A disadvantage of this purification method is the phagocytosis of cell debris during the period necessary for attachment of the cells [7], as could be demonstrated by electron microscopic (e.m.) studies and an increase in the protein content of the seeded cells [8]. The undesired uptake of material may interfere with studies on the kinetics of endocytosis by these cultured Kupffer cells [8], and accentuate lysosomal enzyme activities [7].

SEPARATION OF KUPFFER CELLS BY DENSITY-GRADIENT CENTRIFUGATION

Kupffer and endothelial cells from rat liver have nearly the same density [3] and can only be separated by means of density-gradient centrifugation if the density of one of the cell types has been changed prior to the centrifugation. After preferentially loading lysosomal structures in Kupffer cells *in vivo* with Triton WR 1339 or with an iron-sorbitol-citric acid complex, Jectofer^R, Kupffer cells were separable in Metrizamide [3] (yield of $6-7 \times 10^6$ cells/g rat liver [1, 3]), still with ~20% endothelial cells present [3]. Furthermore, the purified Kupffer cells showed an altered ultrastructure and increased lysosomal enzyme activities [3].

PREPARATION OF KUPFFER CELLS BY CENTRIFUGAL ELUTRIATION

Suspensions obtained by pronase or collagenase treatment can serve as starting material [1, 4, 9]. The equipment for centrifugal elutriation (Beckman Instruments JE-6 Rotor) and its use, depending on countercurrent centrifugation, for the separation of sinusoidal liver cells have been described previously [9]. Highly pure and viable Kupffer cells are obtainable from both rat liver [3, 9] and mouse liver [1]. The yields (maximum 8×10^6 cells per g rat liver [3]), purity and viability are as stated above [1]. Kupffer cell suspensions thus prepared from mouse and rat liver can be directly used for biochemical analyses or cultured for studies on the mechanisms of endocytosis [8].

PREPARATION OF ENDOTHELIAL CELLS BY THESE APPROACHES

Highly purified endothelial cells, 90% viable, are separable from the loaded Kupffer cells in Metrizamide: the band at density

1.064 g/ml or less is 93% pure, and the yield is about 4 × 10⁶ cells/ g liver [3].

By centrifugal elutriation as described for Kupffer cells, endothelial cells of purity and viability at least 90% are obtainable in the above-mentioned yield, or even 28 × 10⁶ cells/g rat liver under favourable conditions [3].

Table 1. Properties of purified rat-liver Kupffer and endothelial cells [with refs.].

	Kupffer cells	Endothelial cells
Diameter, μm	8.7-9.1 [10]	7.0 [11]
Volume, nl/cell	345-394 [10]	179 [11]
Protein, μg/10⁶ cells	78-116 [10], 138 [2], 154 [6]	47 [3]

PROPERTIES OF ISOLATED KUPFFER AND ENDOTHELIAL CELLS

Purified Kupffer cells have about twice the volume and protein content of endothelial cells (Table 1). By scanning e.m., the surface characteristics of isolated Kupffer cells [8, 12-14] and

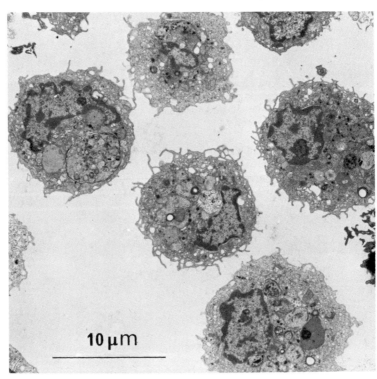

Plate 1. Kupffer cells isolated from a female BN/BiRij rat. The cells were purified by centrifugal elutriation.

10 μm

endothelial cells [12] have been studied. The surface of freshly
isolated Kupffer cells is covered with large lamellipodes [12].
Functional Fc and C$_3$ receptors could be demonstrated on the surface of cultured Kupffer cells [13]. Isolated enothelial cells
are provided with lamellipodes and fillipodia, and numerous
pores can be observed at the cell surface [12].

By transmission e.m. [2, 3, 5, 8, 10, 12], isolated Kupffer
cells (Plate 1) largely resemble Kupffer cells in the intact
liver, the main exception being that no fuzzy coat was observed
on the isolated cells [2]. *In situ,* endothelial cells are thin
cells with cellular processes lining the sinusoidal lumen. After
isolation, these processes with their characteristic sieve
plates are retracted but are still present in the form of a
sponge-like structure in which sieve plates can be recognized
[2, 3, 9, 15] (Plate 2).

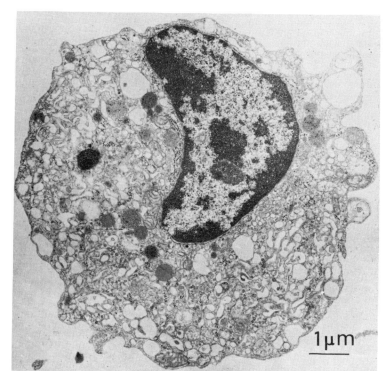

Plate 2. Endo-
thelial cell
isolated from the
liver of a female
BN/BiRij rat and
purified by centri-
fugal elutria-
tion.

1 μm

References

1. Knook, D.L. & Sleyster, E.Ch. (1977) in *Kupffer Cells and Other Liver Sinusoidal Cells* (Wisse, E. & Knook, D.L., eds.), Elsevier/North-Holland, Amsterdam, pp. 273-288.
2. Emeis, J.J. & Planqué, B. (1976) *J. Reticuloendothel. Soc. 20*, 11-29.
3. Knook, D.L., Blansjaar, N. & Sleyster, E.Ch. (1977) *Exp. Cell Res. 109*, 317-329.
4. Roser, B. (1968) *J. Reticuloendothel. Soc. 5*, 455-471.
5. Mills, D.M. & Zucker-Franklin, D. (1969) *Am. J. Pathol. 54*, 147-166.
6. Munthe-Kaas, A.C., Berg, T., Seglen, P.O. & Seljelid, R. (1975) *J. Exp. Med. 141*, 1-10.
7. Berg, T. & Munthe-Kaas, A.C. (1977) *Exp. Cell Res. 109*, 119-125.
8. Brouwer, A. & Knook, D.L. (1977), *as for* [1], pp. 343-352.
9. Knook, D.L. & Sleyster, E.Ch. (1976) *Exp. Cell Res. 99*, 444-449.
10. Sleyster, E.Ch., Westerhuis, F.G. & Knook, D.L. (1977), *as for* [1], pp. 289-298.
11. Knook, D.L. & van Doorn, E., *unpublished results*.
12. Drochmans, P., Sleyster, E.Ch., Penasse, W., Wanson, J.C. & Knook, D.L. (1977), *as for* [1], pp. 131-139.
13. Munthe-Kaas, A.C., Kaplan, G. & Seljelid, R. (1976) *Exp. Cell Res. 103*, 201-212.
14. Polliack, A. & Hershko, C. (1978) *Blood Cells 4*, 301-318.
15. Knook, D.L., Sleyster, E.Ch. & Van Noord, M.J. (1975) in *Cell Impairment in Aging and Development (Advances in Experimental Medicine and Biology,* Vol. 53*)*, Plenum, New York, pp. 155-169.

#B-3
LEUCOCYTE SEPARATION BY SEDIMENTATION AT UNIT GRAVITY

W. S. BONT and J. E. DE VRIES
The Netherlands Cancer Institute,
Plesmanlaan 121,
Amsterdam, The Netherlands

Velocity sedimentation at unit gravity has successfully been used for the separation of cells differing in size [1]. However a serious disadvantage of the existing methods is the relatively long sedimentation time that is required for optimal separations [2]. Furthermore the number of cells that can be separated is limited by the dimensions of the gradient, since the degree separation strongly depends on the thickness of the cell sample layer and the cell concentration of the sample [3]. In general the thickness of the sample layer is reduced by increasing the surface area of the gradient, which results in large gradient volumes. The generation of voluminous gradients is tedious whereas application of the sample becomes increasingly more difficult since large interfaces are easily disturbed [4].

We now describe a new method which facilitates the rapid production of continuous gradients and easy application of the cell sample. The capacity for cell separation is considerably increased, whereas the time required for optimal separation is drastically reduced. The problems encountered with these separations are of theoretical and practical nature. First we will discuss the theoretical aspects and then give a solution for some of the practical problems. The results with the cell separator will be illustrated by the isolation of very pure lymphocytes and highly enriched monocytes from peripheral human blood.

THEORETICAL CONSIDERATIONS

The equation of continuity for the sedimentation of cells at unit gravity in a vessel with a constant cross-section reads

$$\left(\frac{dc}{dt}\right)_h = \left(\frac{dJ_h}{dh}\right)_t \tag{1}$$

where c = concentration of cells, and J_h = no. of cells flowing per sec per cm^2 at a height h;

$$\left(\frac{dc}{dt}\right)_h = \text{change in concentration per sec in a volume}$$

between h and h + dh.

If we consider h_i (the height of the interface between gradient and sample) the following equation holds:

$$\left(\frac{dc}{dt}\right)_{h_i} = \left(\frac{dJ_h}{dh}\right)_t = \frac{V \cdot c_0 - \frac{\eta_0}{\eta} \cdot Vc}{\Delta h} \qquad (2)$$

in which   V = velocity of the cells in the sample,
                  η_0 = viscosity of the sample,
                  c_0 = concentration of cells in the sample,
                  η = viscosity of the gradient,
                  c = concentration of cells in the gradient,
                  h_i = position of the interface,
                  Δh = height of the zone above the interface.

If we replace $- \frac{\eta_0}{\eta} \cdot \frac{V}{\Delta h}$ by P and

$\frac{V \cdot c_0}{\Delta h}$ by Q, we can write (2) as

$$\left(\frac{dc}{dt}\right)_{h_i} = Pc + Q. \qquad (3)$$

A solution for this equation is given by

$$\frac{c}{c_0} = \frac{\eta}{\eta_0} + \left(1 - \frac{\eta}{\eta_0}\right) \cdot e^{\left(\frac{-\eta_0 V}{\eta_0} \cdot t\right)} \qquad (4)$$

It is easy to demonstrate with the aid of equation (4) that in a small region Δh just at the interface the concentration reaches its maximum value

$$\frac{\eta}{\eta_0} \cdot c_0 \qquad \text{within a few minutes.}$$

The viscosity of the sample, η_0, can be raised to that of the gradient without appreciable increase in density, by the addition of a substance with a high intrinsic viscosity, e.g. polyethylene oxide (PEO). The concentration, c, at the interface then always equals the sample concentration, c_0.

The amount of cellular material that can be loaded is determined by the density at the top of the gradient, which can be calculated with the equation

$$d = \left(1 - \frac{d_0}{d_s}\right) \cdot c + d_0$$

where d_0 = density of solvent or suspension medium, d_s = density of solute or suspended particles (the average d_s being 1.065 for mononuclear leukocytes [5]), c = concentration of solute or suspended particles in g/ml.

Thus, the only trivial condition which has to be fulfilled for cell separation procedures at 1 _g_ is that the density of the cell sample must not exceed that at the top of the gradient. The density of a cell suspension is a linear function of the concentration of cellular material. This concentration, c, depends on the volume $V = \frac{4}{3} \pi R^3$ of the cells and their density, according to the relation $c = V \cdot d_s n$, where V = volume of the cell, R = radius, d_s = density of cell and n = number of cells/ml. From this relation it follows that the number of cells that can be loaded differs by about one order of magnitude for cells differing by a factor of 2 in diameter.

Sedimentation processes can be disturbed by bulk sedimentation, a phenomenon called streaming [6]. One type of streaming, which will be termed streaming I, is observed during or directly after layering of a sample when the concentration of cells is raised above a certain maximum value (termed the streaming limit by Miller & Phillips [6]). Streaming I causes the largest disturbance of the sedimentation process and is caused not only by the density of the sample, but also by an increase of density due to a discontinuity in viscosity between sample and gradient as discussed above. However, streaming I can be completely eliminated by raising the viscosity of the sample with PEO (provided that the density of cell sample never exceeds that at the top of the gradient).

A second type of streaming (streaming II) occurs in the gradient, and was originally described for zonal centrifugation in density gradients [7]. However, it also occurs in gradients for sedimentation at unit gravity [1]. Streaming II can be avoided by layering cells in the form of an inverse concentration gradient [7]. However, such a layering procedure is required only if the thickness of the cell sample layer exceeds 0.25 ml/cm² [8] and was omitted in our experiments, since the thickness of this layer in the horizontal position was only 0.07 ml/cm². Streaming II can also be reduced by using relatively low cell concentrations in the sample and by adapting the form of the gradient to the cell load. The larger the cell load, the steeper the gradient has to be, to reduce streaming II [8].

EXPERIMENTAL ASPECTS

Description of the cell separator

The cell separation apparatus (LACS) was purchased from De Koningh BV (P.O. Box 347, Arnhem, Netherlands). It consists of a rectangular perspex sedimentation chamber with inner dimensions 3 × 5 × 60 cm. The chamber can be turned in 30 min from a vertical to a horizontal position and _vice versa_ (Fig. 1).

Rapid layering of discontinuous gradients

Rapid layering of the different solutions of the gradient was
carried out by means of a rectangular metal filling device Fi
(Fig. 2), the bottom of which consisted of a sieve S. Fi was
placed on the bottom of the sedimentation chamber and was pushed
upwards when the first, most dense Ficoll solution was introduced into the chamber *via* a syringe. Fi adhered to the surface
of the gradient as a result of an equilibrium between surface
tension and a weight (W). Fi remained adherent to the surface
of the gradient whilst the gradient solutions were successively
poured into the syringe. Thereby the sieve prevented disturbance of the interfaces (Fig. 3). The cell sample was introduced *via* the same system. The whole procedure took about 15 min,
and could be performed under sterile conditions.

Composition of the gradient

Altogether 12 solutions of
30 ml vol. differing in
density were pre-
pared by dilution
of the following
two stock solutions.
Solution 1 con-
tained 2.5% (w/v)

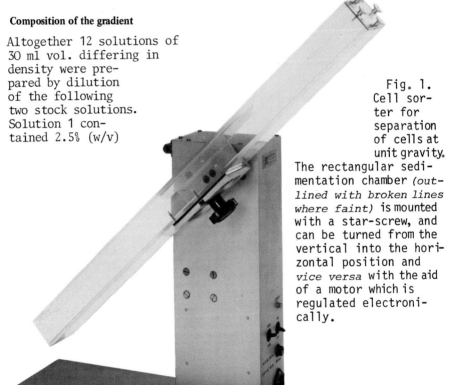

Fig. 1.
Cell sor-
ter for
separation
of cells at
unit gravity.
The rectangular sedi-
mentation chamber *(out-
lined with broken lines
where faint)* is mounted
with a star-screw, and
can be turned from the
vertical into the hori-
zontal position and
vice versa with the aid
of a motor which is
regulated electroni-
cally.

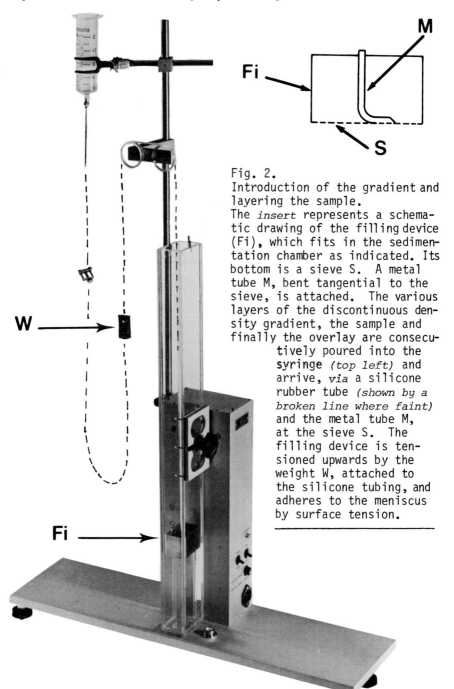

Fig. 2.
Introduction of the gradient and
layering the sample.
The *insert* represents a schematic drawing of the filling device
(Fi), which fits in the sedimentation chamber as indicated. Its
bottom is a sieve S. A metal
tube M, bent tangential to the
sieve, is attached. The various
layers of the discontinuous density gradient, the sample and
finally the overlay are consecu-
tively poured into the
syringe *(top left)* and
arrive, *via* a silicone
rubber tube *(shown by a
broken line where faint)*
and the metal tube M,
at the sieve S. The
filling device is ten-
sioned upwards by the
weight W, attached to
the silicone tubing, and
adheres to the meniscus
by surface tension.

Ficoll, mol. wt. 70,000 (Pharmacia,
Uppsala, Sweden), 0.5% polyethylene
oxide (PEO), mol. wt. 60,000 (BDH,
Poole, U.K. code WSR-205), 2% (v/v)
foetal calf serum (FCS, Flow Laboratories, Ayrshire, Scotland) in
RPMI 1640 medium (Gibco, Grand
Island, N.Y.), buffered to pH 7.2
with 10 mM HEPES (Calbiochem., La
Jolla, Calif.). Solution 2 contained 7.5% (w/v)
Ficoll, mol. wt.
70,000, 2% FCS in
RPMI 1640 buffered
with 10 mM HEPES
to pH 7.2. Both
solutions con-
tained 100 I.U./ml
penicillin, 100 µg/ml
streptomycin and 1 µg/ml
fungizone, and were 300 m-
Osmol. The two stock
solutions were diluted
as follows. Fraction 1,
consisting of 30 ml of
stock solution 2, was
introduced as the first
layer in the sedimenta-
tion chamber. Fraction 2
was prepared by adding
15 ml of stock solution 1
to 150 ml of solution 2.
After mixing 30 ml was
taken out and introduced
into the sedimentation
vessel as the second
layer. The remaining

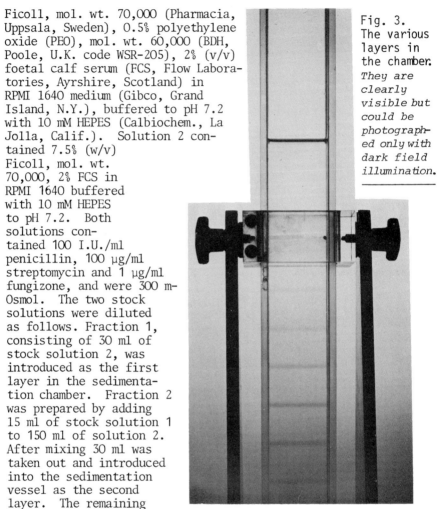

Fig. 3.
The various
layers in
the chamber.
*They are
clearly
visible but
could be
photograph-
ed only with
dark field
illumination.*

135 ml were further diluted with 15 ml of solution 1 and again
30 ml was taken out and introduced as the third layer. This
procedure was repeated another 8 times, and resulted in 11
layers. The twelfth layer consisted of 30 ml of fraction 1. In
this way the same concentration interval between two adjacent
layers was obtained.

Preparation and layering of sample

For the preparation of the cells, mononuclear leucocytes were
isolated from 50-160 ml defibrinated peripheral blood from
healthy donors by centrifugation over a Ficoll-Hypaque mixture
[9]. The cells (50-200 × 10⁶) were washed once with Eagle's
Minimal Essential Medium (EMEM) (Gibco, Grand Island, N.Y.), and

Fig. 4. The tilting
procedure.
The angle of rotation is
plotted against time. The
angular velocity is
slowed down when the
chamber approaches the
horizontal position.

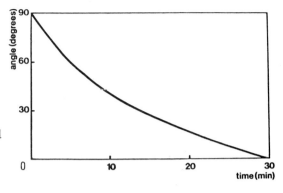

10% FCS and re-suspended
in 25 ml of phosphate-
buffered saline (PBS)
which contained 0.13%
(w/v) PEO. These cell
samples were layered on the gradient and covered with an overlay
of 35 ml PBS by means of the layering device as described above.
After the overlay had been applied on top of the sample the
layering device was removed and the chamber closed.

Rapid transformation of discontinuous into continuous gradients

After closing the sedimentation chamber, it was turned from the
vertical to the horizontal position. The angular velocity,
regulated electronically, slowed down when the vessel reached a
more horizontal position (Fig. 4). Fig. 5 shows that these
manipulations are possible without disturbing the different
layers. The transformation of the discontinuous gradient into
the continuous one did not interfere with the sedimentation process and was completed within 30 min in the horizontal position
(Fig. 6).

Decrease of the sedimentation time

The cells were allowed to sediment while the chamber was in the
horizontal position at 4°. After 2.5 h the chamber was returned
to its vertical position and opened, and its contents fractionated. The chamber has a 20 : 1 ratio of length (60 cm) to width
(3 cm), which implies that the distance between two cells in the
horizontal position is increased by a factor 20 when the chamber
is returned to its vertical position. Therefore the time required for the separation of two cell populations was also decreased by a factor of 20.

Fractionation of the gradient

The fractionation was carried out with a specially designed
syphoning device (Fig. 7). The sedimentation velocity was expressed as a ratio — the distance the cells had sedimented after
the chamber had returned to its vertical position, relative to
the sedimentation time. Fractions with velocity 20-36 mm/h contained only lymphocytes, whereas the cells with sedimentation
velocities in the range 50-62 mm/h were highly enriched in monocytes.
The fractionation was completed within 15 min. The different fractions

were spun 15 min at 800 *g* at 4°,
and the cells were washed twice
by re-suspending in cold EMEM +
1% FCS and spinning at 150 *g* for
10 min at 4°. Finally, the cells
were re-suspended in EMEM + 10%
FCS, counted and used for morphological and functional studies.

Cell characterization

The different mononuclear leucocyte fractions were characterized
according to different criteria:
a) morphologically after May-
Grünwald-Giemsa (MGG) staining of
cytocentrifuge preparations;
b) staining for non-specific
cytoplasmic esterase [10];
c) electronic sizing with a
Coulter counter model ZF supplemented by a pulse-height analyzer
(Chanelyzer model C-1000) [11];
d) phagocytosis of fluorescent
carboxylated beads of average
diameter 1.6 μm (Polysciences,
Warrington, PA), monitored by
fluorescence microscopy [12].

RESULTS

In 6 different experiments
$4\text{-}20 \times 10^7$ mononuclear leucocytes were separated (Table 1).
Approximately 60% of all lymphocytes were recovered in the
first fraction, with the lowest
sedimentation velocity, desig-
nated as *lymphocyte fraction*
(LF) since the average monocyte
contamination was only 0.2% as
judged by staining for non-
specific esterase. Fraction
no. 5, with the highest sedimentation velocity, contained 40 %
of all monocytes, and generally

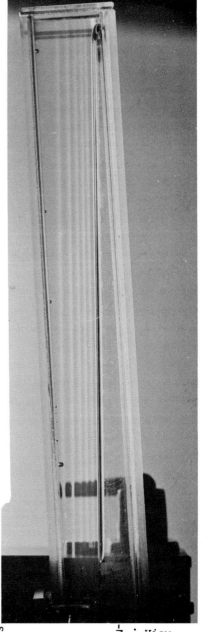

Fig. 5. The layers during the procedure. The dis- continuous gradient in the chamber when it is almost in the hori- zontal position. It is shown that the vari- ous layers were not disturbed by the tilting procedure (photo- graphed with dark field illu- mination). *View Fig. 5 sideways (← TOP).*

Fig. 6. Transformation
of the discontinuous
gradient into a con-
tinuous one.
The refractive index
(—, expected; ● ,
measured) was plotted
against fraction number.
The discontinuous gra-
dient was transformed
into a continuous one
after the chamber had
been in the horizontal
position for 30 min.

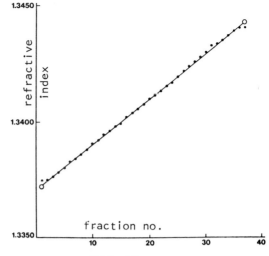

over 80% of the cells
were monocytes; it
was therefore desig-
nated as *monocyte (enriched) fraction* (MF). The average recovery
of all cells after 2 washings was 84 ±5%, whereas the total recovery
of monocytes in all monocyte-containing fractions was 77 ±5%. From
these recoveries it may be concluded that no selective loss of
monocytes occurred during the isolation procedure. The viability
of both lymphocytes and monocytes was 99% as judged by trypan
blue exclusion.

Further characterization of the LF and MF is given in Table 2.
The LF contained 56 ±6% E-rosette-forming cells, which was
slightly lower than expected, whereas the percentages of lymphocytes bearing Fc, C_3 receptors or sIg was not significantly
different from the average percentages of these cells in the
corresponding UL, indicating that no enrichment or depletion of
these cells occurred. The MF contained 16 ±4% contaminating
lymphocytes, consisting of 10% Fc and 5% C_3 receptor-bearing
cells; 3% of the lymphocytes were positive for sIg. Fractions
2-4 containing ∿40% of all lymphocytes and 60% of all monocytes
were not further investigated in this study.

DISCUSSION

Several general aspects of 1 *g* sedimentation procedures[*] for cell
separation determined our strategy for the separation of human
mononuclear blood leucocytes. The resolution obtained after
separation of cells differing in size depends on many criteria
which are inter-related, such as the sedimentation time and the
thickness of the cell sample layer [3], the cell load and the
shape of the gradient [6], the viscosity of the sample layer [4]
and the volume and dimensions of the sedimentation chamber [13].

*Cf. A.M. Denman & B.K. Pelton (lymphocytes) in Vol. 2, *this series.-ED.*

A general disadvantage
of separating cells
at unit gravity is
that compared to
centrifugation procedures,
relatively long sedimen-
tation times are required.
However, centrifugation
techniques can lead to
high cell losses due to
wall effects, particularly
for the most rapidly sedimenting cells [2]. The
selective cell losses can
be prevented by using
special and expensive
rotor types, e.g. zonal
rotors, or elutriator
rotors (cf. D.L. Knook &
E. Ch. Sleyster, this vol.).

Therefore we have
focused our attention
on aspects related to
the time required to
achieve an optimal
degree of cell sepa-
ration at unit gravity.
The sedimentation time
was found to be dras-
tically reduced by the
introduction of the
tilting procedure.
By turning the chamber
from the vertical to
a horizontal position
the surface area was
increased 20-fold and
the thickness of

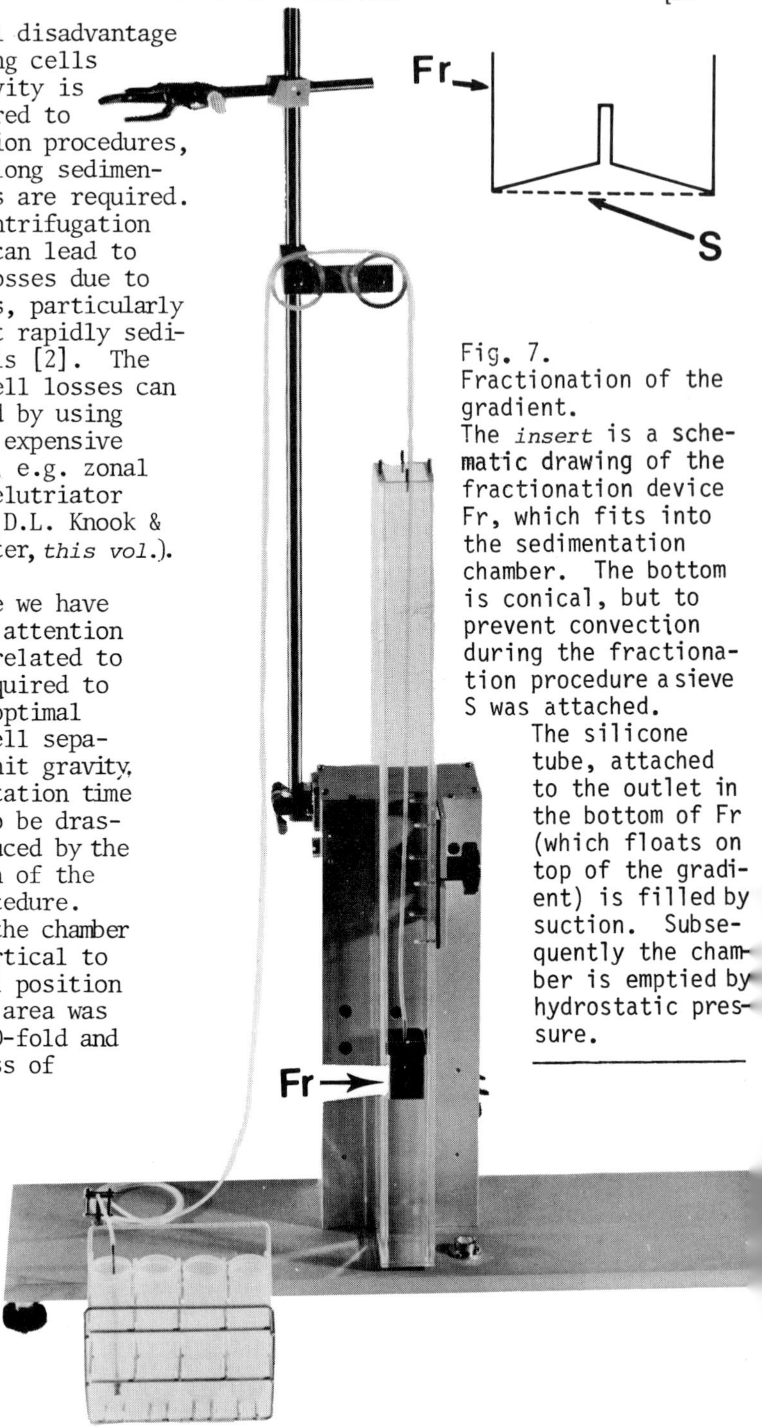

Fig. 7.
Fractionation of the
gradient.
The insert is a sche-
matic drawing of the
fractionation device
Fr, which fits into
the sedimentation
chamber. The bottom
is conical, but to
prevent convection
during the fractiona-
tion procedure a sieve
S was attached.
The silicone
tube, attached
to the outlet in
the bottom of Fr
(which floats on
top of the gradi-
ent) is filled by
suction. Subse-
quently the cham-
ber is emptied by
hydrostatic pres-
sure.

Table 1. Isolation of human blood lymphocytes and monocytes by velocity sedimentation at 1 g.

Expt. no.	Starting material (mononuclear leukocytes)			Lymphocyte fraction (LF)		Monocyte (enriched) fraction (MF)				
	No. of leukocytes[1] ×10⁷	% Monocytes		% Yield[4]	% Viability[5]	% Monocytes			% Yield[7]	% Viability[5]
		morph.[2]	sizing[3]			morph.[2]	sizing[3]	phagocytosis[6]		
1	8.5	22	21	60	99	85	81	81	48	99
2	4.1	16	18	68	99	78	75	73	40	99
3	10.2	26	23	57	98	92	87	85	32	99
4	5.8	31	28	54	98	88	84	77	44	99
5	7.2	18	17	59	98	82	78	83	38	99
6	19.6	20	19	64	98	80	83	ND	36	99
Mean ± S.D.:	—	22 ±5	21 ±4	60 ±5		84 ±5	81 ±4	80 ±5	40 ±6	

1 number of mononuclear cells added to the gradient;

2 counting minimally 300 cells in cytocentrifuge preparations stained for non-specific esterase;

3 electronic sizing (see text);

4 compared to all lymphocytes recovered after fractionation;

5 as judged by trypan-blue exclusion tests counting 300 cells;

6 phagocytosis of fluorescent beads, counting 300 cells;

7 compared to all monocytes recovered after fractionation.

Table 2. Characterization of the unfractionated lymphocytes (UL),
the LF and MF (*cf.* headings in Table 1).
The values represent the mean ± S.D. of 5 different donors.

	UL	LF	MF
Sedimentation velocity, mm/h	-	12-36	50-62
% Monocytes (esterase)[1]	23 ±4	0.2	83 ±4
% Monocytes (phagocytosis)	20 ±5	<0.1	79 5
% Monocytes (electronic sizing)	22 ±4	-2	81 4
E-rosettes	51 ±6	56 ±6	9 ±3
Lymphocytes with F_c receptor[3]	15 ±3	17 ±4	10 ±3
Lymphocytes with C_3 receptor[3]	18 ±4	20 ±3	5 ±2
Lymphocytes with sIg[4]	6 ±1	7 ±2	3 ±1
Granulocytes (basophils)	<0.5	0	1

[1] No. of cells counted was at least 300, but was 1000-2000 for
the monocyte contamination check on the LF.
[2] None detectable.
[3] Lymphocytes with F_c or C_3 receptors in microscopic preparations
stained for non-specific esterase.
[4] Monocytes were recognized by phagocytozed fluorescent beads.

the cell sample layer, which was sandwiched between the overlay
and the top of the gradient, was correspondingly reduced. This
implied that the time required for optimal separation was reduced 20-fold [13]. Returning the chamber to the vertical
position increased the distance between two cells 20-fold. This
gain in resolution was preserved during fractionation, because
the construction of the fractionation devices enabled us to
collect very thin layers from the gradient without convection.
With our modification the capacity for cell separation at unit
gravity, expressed as no. of cells per litre of gradient per hour,
was 10 times higher than described by Brubaker & Evans [14].

As considered above (Theoretical section), streaming II is reduced by starting with very thin sample layers and steep density
gradients [8]. The tilting procedure therefore diminishes
streaming II for the following reasons.

The slope of the density gradient is defined by $\frac{dc}{dh}$, where
c = the concentration of the solute and h = the
height of the gradient column.

Since h was reduced by the tilting procedure the slope of the
gradient was increased. This effect, in combination with the reduction in thickness of cell sample layer (both achieved by the

tilting procedure), minimized streaming II [8]. Finally it was demonstrated experimentally that application of the cells with an inverse concentration gradient on top of the density gradient had no influence on streaming I (i.e. the streaming limit defined by Miller & Phillips [6]).

Leucocyte separation

In spite of the fact that human blood leucocytes are one of the best studied cell populations, data on the fractionation of these cells into various pure sub-populations by physical methods only are scarce (but see, *inter alia*, M.J. Owen & M.J. Crumpton, *this vol.*).Therefore the separation of these cells is given as an example of the potentialities of this apparatus.

Highly enriched monocyte fractions were obtained, generally consisting of more than 80% monocytes. The yield and purity were better than those obtained by a combination of surface adherence and density separation [15] and density separation only [5]*. Moreover, cell separation by density centrifugation has to be performed under carefully standardized conditions in order to avoid irreproducible results, and the number of cells that can be separated in a single experiment is limited [5].

In the same sedimentation runs ∿60% of the lymphocytes were obtained very pure. The monocyte contamination in these LF varied from 0.01 to 0.2% as judged by staining for a non-specific cytoplasmic esterase. Less pure lymphocyte populations resulted from purification procedures based on monocyte depletion by selective removal through adherence to nylon wool or plastic surfaces columns [16, 17], selective destruction by specific antisera and complement [18] or phagocytosis of iron particles and a magnet [19]. Also, preparations of monocytes from lymphocytes with a multi-parameter cell sorter [20] are still too complicated for large-scale isolation of different cell types. Another disadvantage of the monocyte depletion methods mentioned above is that the monocytes are difficult to recover; they are no longer usable for functional studies or are completely destroyed. The functional properties of the monocyte-enriched fractions obtained by velocity sedimentation are not affected because the cells are obtained in suspension (and not by adherence, with leads to activation), which implies that they can easily be recovered and can be used for studies on cell differentiation and function.

The method was also successfully used to separate murine leukaemia cells in different phases of the cell cycle [21], and proliferating myeloid cells from normal and leukaemic human bone marrow [22]. Preliminary data indicate that suppressor cells
*
 Density-dependent separation of haemoietic cells is considered by K.A. Dicke in Vol. 2, *this series.* — *Ed.*

could be separated from peripheral blood-cell preparations of non-Hodgkin lymphoma patients [23]. Through simplified generation of large-volume gradients and reduction of the sedimentation time, this separation method with a larger version of the apparatus has enabled monocytes to be isolated in large quantities. The separation of cells from tissues such as spleen and testis has been successfully performed (unpublished work). [For other 1 g work, see elsewhere in this vol., particularly P.O. Seglen.- Editor.]

References

1. Mel, H.C. (1964) J. Theor. Biol. 6, 181-200.
2. Pretlow, T.G., Weir, E.E. & Zettergren, J.G. (1975) Int. Rev. Exp. Pathol. 14, 91-204.
3. Peterson, E.A. & Evans, W.H. (1967) Nature (Lond.) 214, 824-825.
4. Tulp, A. & Bont, W.S. (1975) Anal. Biochem. 67, 11-21.
5. Loos, J.A., Blok-Schut, B., Kipp, B., Van Doorn, R., Hoksbergen, R., Brutel de la Rivière, A. & Meerhof, L. (1976) Blood 48, 731-742.
6. Miller, R.G. & Phillips, R.A. (1969) J. Cell Physiol. 73, 191-201.
7. Britten, R.J. & Roberts, R.B. (1960) Science 131, 32-33.
8. Noll, H. (1969) in Techniques in Protein Biosynthesis (Cambell, P.N. & Sargent, J.R., eds.), Academic Press, London, pp. 101-179.
9. Böyum, A. (1968) Scand. J. Clin. Lab. Invest. 21, suppl. 97, 77-89.
10. Yam, L.T., Li, C.Y. & Crosby, W.H. (1971) Amer. J. Clin. Pathol. 55, 283-290.
11. Loos, J.A., Blok-Schut, B., Kipp, B., Van Doorn, R. & Meerhof, L. (1976) Blood 48, 743-753.
12. Cline, M.J. & Lehrer, R.I. (1968) Blood 32, 423-435.
13. Bont, W.S. & de Vries, J.E. (1977) in Cell Separation Methods (Bloemendal, H., ed.), N. Holland, Amsterdam, pp. 3-13.
14. Brubaker, L.H. & Evans, W.H. (1969) J. Lab. Clin. Med. 73, 1036-1041.
15. Brodersen, M.P. & Curns, C.P. (1973) Proc. Soc. Exp. Biol. Med. 144, 941-944.
16. de Vries, J.E. & Rümke, P. (1976) Int. J. Cancer 17, 182-190.
17. Bean, M.A., Bloom, B.R., Herbermann, R.B., Old, L.J., Oettgen, H.F., Klein, G. & Terry, W.D. (1975) Cancer Res. 35, 2902-2913.
18. Greaves, M.F., Falk, J.A. & Falk, R.E. (1975) Scand. J. Immunol. 4, 555-562.
19. Holm, G.P., Petterson, D., Mellstedt, H., Hedfors, E. & Bloth, B. (1975) Clin. Exp. Immunol. 20, 443-457.
20. Kwan, D., Epstein, M.B. & Norman, A. (1976) J. Histochem. Cytochem 24, 355-362.
21. Bont, W.S. & Hilgers, J. (1976) Prep. Biochem. 7, 45-60.
22. Burghouts, J., Plas, A.M., Wessels, J., Hillen, H., Steenbergen, J. & Haanen, C. (1978) Blood 51, 9-21.
23. Vyth, F. (1979) Unpublished experiments.

#B-4

CELL SEPARATIONS IN A NEW DENSITY GRADIENT MEDIUM, PERCOLL®

HÅKAN PERTOFT[1], MICHAEL HIRTENSTEIN[2] and
LENNART KÅGEDAL[2]
[1]Institute of Medical and Physiological Chemistry,
University of Uppsala, Uppsala, Sweden,
and [2]Pharmacia Fine Chemicals AB, Uppsala, Sweden.

Percoll^R is composed of colloidal silica coated with poly-
vinylpyrrolidone (PVP). The combination of low viscosity, low
osmolality and stability in physiological saline has been shown
to be advantageous for many cell separations on the basis of
size and/or density. Gradients of Percoll may be formed using
conventional gradient makers, or self-generated in angle-head
rotors. On pre-formed gradients, cells band at their isopycnic
densities in 10-20 min at low g forces. Examples of separations
are given.—
(a) Rat liver cells can furnish sub-populations of parenchymal
cells (hepatocytes) having buoyant densities of 1.07-1.09 g/ml,
and non-parenchymal cells (mostly phagocytosing Kupffer cells)
at a density of 1.04-1.06 g/ml. Non-viable cells accumulate
at the top of the gradient. The cells recovered from the gradient
have been maintained in culture with intact physiological activity.
(b) Macrophages from rat peritoneal fluid and monocytes from
human peripheral blood have also been isolated on Percoll gradients. Lymphocytes from tonsillar tissue and human peripheral
blood have been sub-fractionated on pre-formed gradients of
Percoll, on the basis of both size and density.
(c) Pancreatic islets have been isolated by velocity sedimentation at unit gravity, and α- and β-cells from the islets can be
segregated by density difference.

For isolation of intact, viable, functionally specific cells
from whole tissues, density gradient centrifugation is a commonly used method, facilitating rapid separation on the basis of
density and/or size. The success of cell separations by centrifugation is usually limited by the characteristics of the gradient medium used. The various gradient media cited in the
literature [see G.C. Hartman et al. in Vol. 4, this series—Ed.],
e.g. sucrose, serum albumin, Ficoll^R, Metrizamide, heavy salts,
Ludox etc., have all found applications in specific areas, but
have shortcomings at the concentrations required for cell

separations. Ideally, a gradient medium should be iso-osmotic, non-toxic, have physiological ionic strength and pH, have low viscosity, be capable of forming gradients of 1.0-1.3 g/ml, and be easily removable from biological material.

Suspensions of colloidal silica (Ludox) were first used for density centrifugation by Mateyko & Kopac [1], and later were systematically evaluated by Pertoft and co-workers [2-4]. Unmodified silica sols are unstable in salt solutions and are toxic to cells. It was found that supplementation of the sols with polymers such as dextran, polyethylene glycol or poly-vinylpyrrolidone (PVP) decreased the toxicity of such colloidal silica media. However, the concentration of polymers required increased the viscosity of the medium and made the colloid less stable, and the polymer was difficult to remove from isolated cell preparations.

PHYSICAL PROPERTIES OF 'PERCOLL'

Modified colloidal silica (Percoll^R; Pharmacia Fine Chemicals, Uppsala, Sweden) was recently introduced as a density medium [4, 5] with many superior properties compared with conventional materials and Ludox-PVP mixtures. Percoll consists of colloidal silica particles of 10-30 nm diameter with a PVP coat corresponding to a monomolecular layer [5]. Whereas Ludox-PVP mixtures require a high free polymer concentration to obtain a coating on the silica, Percoll has virtually no free PVP, and the PVP coat on the silica particles cannot be removed by physical means. Gel chromatography of Percoll on Sepharose^R 4B shows a single homogenous peak containing both silica and PVP and a minimal peak of free PVP [5]. In contrast to regular mixtures of silica sols and polymers, Percoll particles show little aggregation in saline solution (Plate 1). They have a lower charge than pure silica particles. Percoll also possesses rather different osmolar and viscosity properties from Ludox, as shown in Fig. 1 together with some values for other media.* Percoll, which can form gradients in the range 1.0-1.3 g/ml, has a unique combination of properties including lack of toxicity [4]. Whereas either osmolality or viscosity become limiting even at fairly low densities with other media, the very low osmotic pressure of Percoll facilitates the formation of gradients which are virtually iso-osmotic throughout, and of sufficiently low viscosity to allow very rapid isopycnic separation of cells. Perhaps the greatest advantage of Percoll as a centrifugation medium is its ability to form gradients *in situ* in 10-20 min, as described below.

---

*   Properties are also given by P.O. Seglen (Table 5), *this vol.*, and its structure is shown diagrammatically by E.M. Reardon *et al., this vol.*

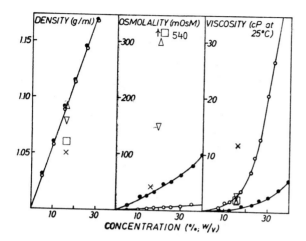

Fig. 1. Physico-
chemical properties
of unsubstituted
colloidal silica
(Ludox HS) ●——●
and Percoll, ○——○.
Additionally, single
values are given
for other media:
sucrose, □ ;
sodium metrizoate,
Δ; metrizamide,▽;
and Ficoll, x.
*From ref. [4], by
kind permission of
the authors and
Plenum Publishing
Co.; methods given
therein. (SERIES
EDITOR'S NOTE: for
other media see
Vol. 4 of this
series — G.C. Hart-
man et al.)*

METHODOLOGY

Preparation of gradient material

Percoll is first made iso-osmotic with the cell preparation. This is easily achieved by adding 1 part (by vol.) of concentrated (× 10) physiological salt solution to 9 parts of Percoll. The isotonic Percoll solution may then be diluted to a lower density by adding isotonic salt solutions.

Gradient formation

Pre-formed density gradients of isotonic Percoll can be made by conventional means, up to d = 1.12 g/ml. Discontinuous gradients, formed by placing isotonic Percoll layers of different densities on top of one other, are useful for low speed or 1 g separations, collecting material at the interfaces [4].

Self-generated gradients are formed when Percoll solutions are centrifuged in angle-head rotors at >10,000 g. Fig. 2 shows a typical gradient formed when Percoll in 0.15 M salt solution starting at d = 1.07 g/ml was centrifuged at 17,500 g in a rotor having a fixed angle of 29°. The 'S'-shaped gradient forms approximately isometrically on either side of the starting density, and becomes progressively steeper with time. Self-generating gradients arise because the colloidal silica particles sediment at high values of g and, since they are of different sizes, the larger particles sediment faster than small ones, creating a concentration gradient and thus a density gradient. An increased proportion of larger particles has actually been demonstrated [6] at the dense end of a self-generated gradient of Percoll.

Plate 1. Elec-
tron microscopy
of Percoll par-
ticles. Negative
contrast with 1%
uranyl acetate
at pH 4.6.

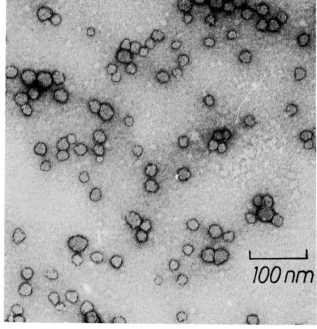

100 nm

The rate of gradient formation and the shape attained are
affected by the g value applied, the centrifugation time (Fig. 2),
the ionic strength of the medium [4], and the angle of the
rotor (Fig. 3). Amongst the methods for measuring density and
gradient shape [4, 5], the most convenient is to use 'density
marker beads' as mentioned in the legend to Fig. 2: these are
made from modified Sephadex[R] and are now commercially available;
Rickwood [7] has drawn attention to their value.

Detection and assay of biological material

Fractions may be collected by standard techniques [4], and cells
detected by using a flow-through cuvette to monitor the turbidity.
Since PVP absorbs UV light [4], quantitative turbidity measurements in the UV-region call for accurate blanking with equivalent
concentrations of colloid, as considered elsewhere [4].

As Percoll has no effect (Fig. 4) on protein determinations
by the Lowry method [8], these can be carried out on fractions
without prior removal of gradient material, as can enzyme assays;
in some instances enzymes are stabilized compared with equivalent
assays carried out in isotonic sucrose [4] [see also 9].

Percoll does not penetrate into cells or adhere to biological
membranes. This is evident from the following values (as counts/
min) obtained with [125I]-labelled Percoll during the isolation
of rat liver hepatocytes (at d = 1.07-1.09):

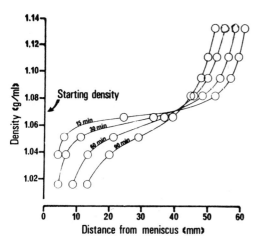

Fig. 2. Time course of
development of Percoll
gradients during centri-
fugation *(see text)* in an
8 × 14 ml angle rotor
(MSE Superspeed 75
centrifuge). Density
values were assessed
by means of coloured
'density marker beads',
DMB (Pharmacia Fine Chemicals AB, Uppsala, Sweden).
The density of each grade
depends on the medium.

^{125}I counts

— 5 ml of cell suspension in Percoll   35,680
— Cell pellet washed with 80 ml of Eagle's MEM   71
— Washing repeated once ....   0.
— Cells from 2 ml of the cell suspension were seeded on
  a Petri dish (80% of the cells attached) and after 4
  washings with Eagle's MEM, the cells were detached
  with 0.01% trypsin + 0.025% EDTA   0

Osmolality

The influence of osmolality on the apparent buoyant density of
particles has been demonstrated using cells, whole organs *(see
below)* and subcellular particles [4, 7]. The increase in buoyant
density with increasing osmolality was confirmed with rat hepatocytes, which band at 1.06-1.12 g/ml depending on the medium
(Fig. 5).

APPLICATIONS, ESPECIALLY FOR CELLS

Separation of liver cells

In iso-osmotic gradients, excellent separations of hepatocytes
and Kupffer cells can be obtained (Fig. 6). The two types of
cells were identified by their specific uptake of asialocerulo-
plasmin and aggregated albumin respectively and, cytochemically
[12], by specific esterase staining. The two cell populations
were thus shown to be completely separated from each other. The
centrifugation technique has also been used to demonstrate that
intravenously injected chondroitin sulphate is taken up preferentially by Kupffer cells in the liver (Fig. 7). Rubin *et al.* [14]

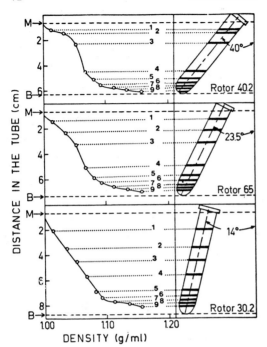

Fig. 3. Rotor angle in
relation to development
of density gradients
during centrifugation of
Percoll (Beckman rotors).
The starting density of
Percoll (0.15 M with
respect to NaCl) was
1.065 g/ml. Centrifuga-
tions were performed at
35,000 g_{av} for 15 min,
at 20°. M and B denote
menisci and bottoms of
the tubes. (The numbers
indicate the banding of
marker beads: *see* legend
to Fig. 2.)

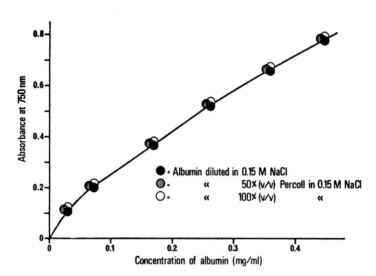

Fig. 4. Colour development of albumin in the Lowry
reaction [8] performed in Percoll of different concentrations. Blank values have been subtracted.

Fig. 5. Frac-
tionation of
rat liver
parenchymal
cells (35 ×
10^6 cells in
2 ml) on a
self-generated
Percoll gra-
dient (8 ml
solution,
d = 1.065 g/ml).
The osmolality
of the Percoll
solution was
varied by
addition of
NaCl to 200
mOsm, 300
mOsm and 400
mOsm. Centri-
fugation was
performed in
a Beckman
rotor 30.2
at g_{av} =
35,000 for
15 min at
4°.

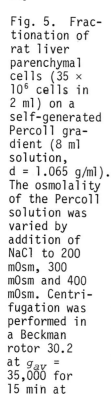
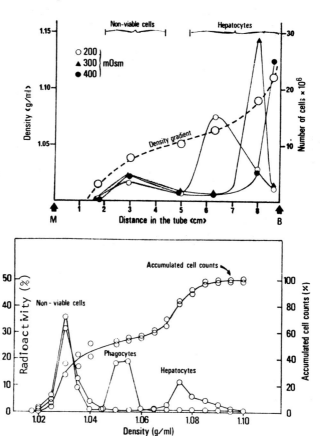

Fig. 6. Separation of various rat liver cells on Percoll gradients. [^{125}I]-labelled heat-denatured albumin (21 µCi;
10 mg) was injected intravenously into a rat 30 min before sacri-
fice. The cells were isolated by a perfusion method with collagenase [10] and diluted to a final concentration of 2 × 10^6 cells/
ml in Eagle's MEM. The cells were layered on pre-formed gradients of Percoll (1.03-1.10 g/ml). The sample vol. was 15 ml on
80 ml of gradient solution. Tubes were centrifuged for 30 min
at 800 g in a swing-out rotor at 4°. Radioactivity was measured
in each fraction, and the percentage of the total amount recovered in each fraction (o——o) was plotted vs. density. The
peak at 1.04-1.06 g/ml represents Kupffer cells. The accumulated
number of cells (o——o) was also plotted against density. In an
identical experiment [^{125}I]-asialoceruloplasmin (2.1 µCi; 10 mg)
was injected to identify hepatocytes; the recovery of [^{125}I] in
this experiment is also given (△——△). The hepatocytes band at
1.07-1.09 g/ml. The peak at 1.03 g/ml represents non-viable
cells. (From ref. [11], by kind permission.)

Fig. 7. Uptake of
chondroitin sul-
phate (CS) into
liver cells. [³H]CS
(5 × 10⁶ cts/min in
0.5 ml) was injected
into the tail vein
of a rat. After
15 min the liver was
perfused with colla-
genase, and the cells
were harvested. A
15 ml suspension
containing 13.5 ×
10⁶ hepatocytes and
5.7 × 10⁶ phagocytes
was layered on top

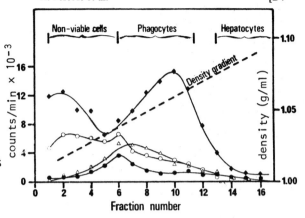

of 75 ml of a pre-formed 10% to 80% (v/v) isotonic Percoll gradient and centrifuged at 800 g for 30 min in a swing-out rotor
at 4°. Fractions were collected and analyzed for density (---)
and radioactivity (o——o). In similar experiments, cells obtained after collagenase and pronase treatment (33 × 10⁶ hepatocytes) were analyzed for the distribution of radioactivity
(◆——◆), purified phagocytes after differential centrifugation
(18 × 10⁶ cells; △——△) and purified hepatocytes after differential centrifugation (35 × 10⁶ cells; ●——●). *The experiment
shows that CS is taken up in the phagocytic fractions. Analysis
of the polysaccharide by gel chromatography on Sephadex G-200
confirmed that the phagocytic fractions also degraded the CS [13].
Note that no radioactivity was found in hepatocytes, i.e. in
cells banding at d >1.07.*

---

have thereby separated cells from antisera during the identification of specific antigens on the hepatocyte surface (Fig. 8).

Blood-cell separations

Fractionation of blood cells has been carried out on pre-formed
(self-generated) gradients of Percoll in two steps, starting
with centrifugation for 5 min at 400 g, during which time the
very small thrombocytes did not penetrate into the gradient.
After removing the plasma with the thrombocytes, and centrifugation for 15 min at 800 g, the other cell types could be banded at
their buoyant densities (Fig. 9).

Lymphocytes have been sub-fractionated isopycnically. Fig.10
shows banding of mouse spleen cells in two apparent fractions in
the density range 1.04-1.09 g/ml. B-cells are present in the
less dense fractions, and the dense fraction presumably contains
T-cells as confirmed by Kurnic *et al.* [15]. When mitogen-
stimulated tonsillar cultures were centrifuged on a continuous

Fig. 8. Demonstra-
tion that anti-
bodies towards
plasma membranes
(p.m.) bind to the
purified hepato-
cytes after incu-
bation of 2 × 10⁶
cells with 15 μl
of various anti-
sera and washing
of the cells,
10 μl of [¹²⁵I]-
labelled Protein A,
which binds to IgG,
was added, and the
cells (15 ml) were
layered on top of

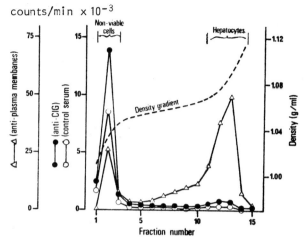

a 50% (v/v) isotonic Percoll solution. The gradient was generated by centrifugation at 25,000 rev/min (40,000 g) for 20 min
at 4° in 100 ml tubes in the MSE 25 centrifuge. After centrifugation, the radioactivity in each fraction was determined.
Antiserum against rat-liver p.m., Δ; antiserum against human
cold-insoluble globulin (CIG), ●; normal rabbit serum, o. *Antibodies against p.m. bind to viable hepatocytes, but only trace
amounts of cold-insoluble globulin could be detected on the
cells [14].*

---

gradient, the highest thymidine incorporation was in a prominent
band in the upper portion of the gradient, whereas small dense
cells were found in the lower part. It was thus possible to
separate blast cells from resting cells [16].

Pure preparations of monocytes have been obtained in blood-
cell fractionations, both on continuous and on discontinuous
gradients: the macrophages consistently band at d = 1.06-1.07
g/ml. The identification of these cells as macrophages was confirmed by esterase activity [12] and by their rosetting behaviour
with complement-treated (C3b) sheep erythrocytes (Plate 2).
Macrophages from different tissues appear to have buoyant densities in the range 1.03-1.07 g/ml when banded on gradients of
Percoll (Table 1). Culture of macrophages after isolation on
Percoll produces fully active, viable cells.

Pancreatic islets of Langerhans

Past approaches are surveyed elsewhere (G. Raydt, *this vol.*).
Buitrago *et al.* [23] have reported rapid isolation of pancreatic
islets by 1-g sedimentation through a step gradient of Percoll
(Plate 3). Islets thus isolated are fully viable and may be

Table 1. Buoyant densities of cells in Percoll gradients. For
cells other than macrophages, the listing is in approximate
order of densities. *Ref. denoted '-' if present work (H.P.).*

Cell type		Buoyant density, g/ml	Ref.
Macrophages	Human blood monocytes	1.06-1.07	18
	Mouse peritoneal macrophages	1.06-1.07	-
	Rat peritoneal macrophages	1.03-1.04	-
	Kupffer cells from rat liver	1.04-1.06	11
Rat epididymal fat pad cells (pre-adipo-cytes)		1.02-1.04	19
Primary calf testicular cells		1.03-1.06	4
HeLa cells from spinner culture		1.04-1.07	11
Microalgae of marine organisms		1.04-1.12	20
Mouse Leydig cells		1.05-1.07	21
Lymphocytes from mouse spleen		1.05-1.10	4
Lymphocytes from mouse peritoneal fluid		1.06-1.09	-
Mast cells from rat peritoneal fluid		1.09-1.13	-
Yeast cells *(Saccharomyces cerevisiae)*		1.10-1.14	22

transplanted directly into experimental animals without removal
of gradient material [24]. No antigenic response to injection
of Percoll has been found. Separation of α- and β-cells
from guinea-pig pancreatic islets has been achieved in a 40%
Percoll/60% Hank's self-generated gradient (20 min, 20,000 g):
α-cells were found to band at d = 1.04 and β-cells at d = 1.07
g/ml [25].

Other cell types

Gradients of Percoll, generated by centrifugation or with gradient
mixers, have been successfully used to separate many other cell
types, e.g. marine organisms, mast cells, yeast cells for synchronous culture, and primary cells for tissue culture. Table 1
summarizes the densities of cells so far isolated. The advantages of using Percoll are not only the simplicity of methodology
and ease of use, but also the mild conditions and closeness to
physiological conditions which may be achieved. Cells isolated
on gradients of Percoll remain viable, and subsequent culture of
isolated cells may be carried out on glass or by microcarrier
techniques producing colonies having full physiological integrity
and biological competence.

Fig. 9. Separa-
tion of human
blood cells in
Percoll density
gradients.
MNC, mononuc-
lear cells;
PMNC, poly-
morphonuclear
cells; RBC,
erythrocytes.
The tubes were
filled with
10 ml of 70%
(v/v) Percoll
in 0.15 M NaCl
and centri-
fuged in a
Beckman rotor
30.2 (14° angle)

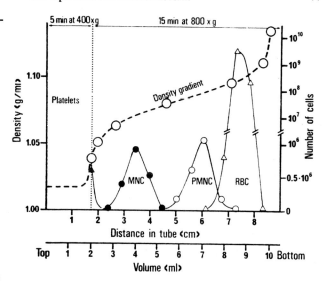

at 20,000 g_{av} for 15 min at 20°. Then 2 ml of gradient solution was removed from the bottom of the tube with a syringe, and
1 ml heparinized blood, diluted with 1 ml 0.15 M NaCl, was
layered on top of the tube. Centrifugation was performed as
indicated. Density was measured by use of 'density marker beads'
in a control tube.

Whole organs

The finding [26] that the density of tissues such as human parathyroid gland may be accurately determined in Percoll gradients
opens up a completely new potential area for research. The proportion of different cell types within a tissue affects the buoyant
density, and this may lead to the use of this technique as a
powerful tool to aid diagnosis of abnormal tissues. Pieces of
parathyroid gland attain their 'true' buoyant density position
when dropped into a column containing a Percoll gradient from
1.00 to 1.08 g/ml. The effect of gradient osmolality on the
apparent density has also been studied in this system: with
Percoll, in contrast with sucrose or Ficoll (not iso-osmotic),
there is negligible 'osmotic drift', i.e. change in density with
time. Percoll is the only convenient gradient material available
which is suitable for such studies, because it can be made iso-
osmotic in a single step irrespective of density.

Subcellular elements

Plant organelles, notably chloroplasts, have been effectively
isolated by the silica-sol approach (see C.A. Price et al. [20]
and C. Jackson et al. [9]. Various components of animal tissue
cells have been studied with the aid of Percoll [4], including

Fig. 10. The density
distribution of cells
from mouse spleen.
For each density level,
the proportion of total
cells is indicated ,
as % (●——●), and of
cells (%) reactive
towards [^{131}I]-labelled
rabbit antibodies
against mouse immuno-
globulin (B-cells)
(o- --o). Cells
labelled with anti-

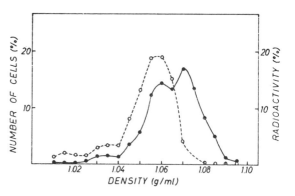

bodies were suspended in Percoll plus Eagle's MEM (d = 1.07
g/ml) to a final cell concentration of 1 × 10^6/ml. Density
gradient centrifugation was performed at 10,000 rev/min for
15 min in a Beckman rotor 65 at 20°. After separation the tube
contents were collected in 0.5 ml fractions, the number of cells
were counted, and the radioactivity was measured by liquid scin-
tillation. *(Source and acknowledgement as for Fig. 1.)*

TEXT, *continued*————————————————

lysosomes [27], plasma membranes [28] and viruses (equine abor-
tion; influenza).

Acknowledgements

We thank Professor T.C. Laurent and Dr. D. Low for helpful
comments. This work was supported by the Swedish Medical
Research Council (Project 13X-4).

Plate 2 *(next page, top)*. Adherence of C3b-coated sheep erythrocytes (E) to purified monocytes. C3b-coated erythrocytes
were prepared by incubation of rabbit anti-E IgM with mouse
serum (C5-deficient AKR mice) for 30 min at 37°[17]. Monocytes
were isolated on isotonic Percoll d = 1.065 g/ml after centrifugation for 15 min at 400 g_{av} in a swing-out rotor at 4°. The
cells were washed in Eagle's MEM and cultivated for 6 days in
monolayers. Cells were re-washed (3 × 5 ml) and the dishes were
covered with C3b-coated sheep erythrocytes for 60 min at 37°.
Photographs x 400. *The erythrocytes have become bound to the
monocytes, demonstrating their complement receptor [18].*

Plate 3 *(next page, bottom)*. Isolation of pancreatic islets.
Collagenase-digested ob/ob mouse pancreas was placed on top of
isotonic Percoll (d = 1.045 g/ml) and allowed to sediment for
10 min at unit gravity. Pure islets were recovered at the
interface to a 'high-density medium' (stock solution of Percoll). Up to 400 islets can be obtained from each mouse pancreas. *From [23], by kind permission.*

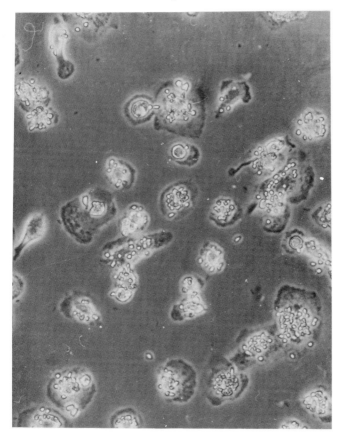

Plate 2
*(Legend on
previous page).*

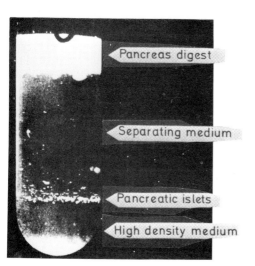

Pancreas digest

Separating medium

Pancreatic islets

High density medium

Plate 3 *(Legend on
previous page).*

References

1. Mateyko, G.M. & Kopac, M.J. (1963) *Ann. N.Y. Acad. Sci. 105*, 185-286.
2. Pertoft, H. & Laurent, T.C. (1968) in *Modern Separation Methods of Macromolecules and Particles*, Vol. 2 (Gerritsen, T., ed.), Wiley Interscience, New York, pp. 71-90.
3. Wolff, D.A. (1975) in *Methods in Cell Biology*, Vol. 10 (Prescott, D.M., ed.), Academic Press, New York, pp. 85-104.
4. Pertoft, H. & Laurent, T.C. (1977) in *Methods of Cell Separation*, Vol. 1 (Catsimpoolas, N., ed.), Plenum, New York, pp. 25-65.
5. Pertoft, H., Laurent, T.C., Lââs, T. & Kågedal, L. (1978) *Anal. Biochem. 88*, 271-282.
6. Pertoft, H. & Laurent, T.C. (1978), *unpublished expts.*
7. Rickwood, D. (1978) in *Centrifugation: a Practical Approach* (Rickwood, D., ed.), Information Retrieval, London, pp. 135-142.
8. Lowry, O.H., Rosebrough, N.J., Farr, A.L. & Randall, R.J. (1951) *J. Biol. Chem. 193*, 265-275.
9. Jackson, C. & Moore, A.L., *this vol.; cf.* Moore, A.L. *et al.* (1977) *Biochim. Biophys. Res. Comm. 78*, 483-491.
10. Rubin, K., Kjellén, L. & Öbrink, B. (1977) *Exp. Cell Res. 109*, 413-422.
11. Pertoft, H., Rubin, K., Kjellén, L., Laurent, T.C. & Klingeborn, B. (1977) *Exp. Cell Res. 110*, 449-457.
12. Barka, T. & Anderson, P.J. (1962) *J. Histochem. Cytochem. 10*, 741-753.
13. Kjellén, L., Pertoft, H. & Höök, M. (1978) *Uppsala J. Med. Sci. 82*, 137.
14. Rubin, K., Oldberg, Å., Höök, M. & Öbrink, B. (1978) *Exp. Cell Res. 117*, 165-177.
15. Kurnick, J.T. and colleagues (1978) *pers. comm.*
16. Kurnick, J.T., Stegagno, M., Sjöberg, O., Örn, A. & Wigzell, H. (1978) *Proc. 12th Internat. Leukocyte Culture Conference, Israel;* p. 42.
17. Bianco, C., Griffin, F.M. & Silverstein, S.C. (1975) *J. Exp. Med. 141*, 1278-1290.
18. Seljelid, R. & Pertoft, H. (1978) *unpublished expts.*
19. Björntorp, P., Karlsson, M., Pertoft, H., Pettersson, P., Sjöström, L. & Smith, U. (1978) *J. Lipid Res. 19*, 316-324.
20. Price, C.A., Reardon, E.M. & Guillard, R.R.L. (1978) *J. Limnol. Ocean. 23*, 548-553 (and Reardon, E.M., *et al., this vol.*)
21. Schumacher, M., Schäfer, G., Holstein, A.F. & Hitz, H. (1978) *FEBS Lett. 91*, 333-338.
22. Haff, L.A. (1977) *Unpublished work (at Pharmacia Fine Chem'ls).*
23. Buitrago, A., Gylfe, E., Henriksson, C. & Pertoft, H. (1977) *Biochem. Biophys. Res. Comm. 79*, 823-828.
24. Henriksson, C. (1978) *pers. comm.*
25. Andersson, T. (1978) *unpublished expts.*
26. Åkerström, G., Pertoft, H., Grimelius, L. & Johansson, H. (1979) *Acta Pathol. Microbiol. Scand., in press.*
27. Pertoft, H., Wärmegård, B. & Höök, M. (1978) *Biochem. J. 174*, 309-317.
28. Öbrink, B., Wärmegård, B. & Pertoft, H. (1977) *Biochem. Biophys. Res. Comm. 77*, 665-670.

#B-5
PARTITION OF CELLS BETWEEN AQUEOUS PHASES

EVA ERIKSSON and GÖTE JOHANSSON
Department of Biochemistry 1,
Chemical Center,
University of Lund, Sweden

Separation of particles by use of partition in aqueous liquid-liquid biphasic systems is surveyed, with attention to the preparation of these systems as well as the parameters driving the partition of particles and their optimization. Rules and experimental guidance are given for affecting the partition by using salts, charged polymers, polymers bearing hydrophobic groups (hydrophobic partition) or groups with specific affinity (affinity partition). Examples of the separation of cells and subcellular elements are cited.

With biphasic systems where particles differ in distribution between two aqueous liquid phases and, in the case of membranes, the interface [1], cells are separated on the basis of their surface properties rather than other factors such as size and particle density. Suitable systems are obtained by mixing aqueous solutions of two polymers, commonly polyethylene glycol (PEG) and dextran. The concentration of PEG in the phase systems is far below that required for cell fusion. Both phases have a high content of water, 80-95%, and are therefore excellent media for biological materials. Salts, sucrose or other water-soluble compounds can be included to adjust tonicity, pH or ionic strength. The technique can easily be scaled up or, as in determining partition behaviour, greatly reduced in scale [2], down to 0.1 ml.

The present account aims to complement that given earlier in this series [3], rather than to re-trace the same ground. To overcome initial problems when working with biphasic systems we start with some practical advice.

PREPARATION OF PHASE SYSTEMS

Choose a phase system of a certain composition. The temperature, tonicity, salt composition and pH should be suitable for the material to be partitioned. An example of a specific system (the % values being w/w) is the following: a phase system (5 g) at +4° containing 5% dextran of mol. wt. 500,000, 4% PEG of mol. wt. 6,000, 0.05 M NaCl, 0.2 M sucrose and 0.01 M pH 7 sodium phosphate buffer. Prepare stock solutions (the first two to be stored in the cold).—

1) 20% dextran. Dextran powder normally contains 3-5% water and therefore the concentration of the solution must be determined. Layer 250 g of dextran on 750 g of water. Heat with gentle stirring and keep the mixture boiling until all the dextran is dissolved. This procedure also minimizes microbiological contamination. To determine the dextran concentration, dilute 5 g of the solution to 25 ml with water and measure the optical rotation of this solution. The specific rotation (for 1 dm, 1 g/ml) is $[\alpha]_D^{25} = + 199$ degrees. Adjust the concentration of the stock solution to 20% with water.

2) 40% PEG. This polymer contains only traces of water. Mix 400 g of PEG with 600 g of water at room temperature. Mechanical stirring is advisable.

3) Salt and buffer solutions. Prepare these 10 times stronger than the final concentrations to be used in the phase systems.

Example.— The amounts of stock solutions necessary for the phase system given above are :— 20% dextran, 1.25g; 40% PEG, 0.5 g: 1 M NaCl, 0.2 ml; 0.2 M Na phosphate buffer, 0.2 ml; 1 M sucrose, 0.8 ml. Mix polymer and salt solutions in a test tube. Add water to 4 g total wt. and bring the tubes to the desired temp. Suspend a sample of the particulate test material in a solution containing 0.05 M NaCl, 0.2 M sucrose and 0.01 M Na phosphate buffer. Add 1 g of this suspension to the phase system. Mix the phases by several (10-40) inversions and allow them to settle. Brief centrifugation of the tubes at very low speed can shorten the separation time but has to be done with care since the sedimentation of the particles is increased. Keep the tubes at a constant temp. during the whole procedure. To determine the partition, withdraw an aliquot from each phase with a pipette and determine the percentage of material in the phases. The amount at the interface is taken as the difference between added material and material in the two phases.

SALT-DRIVEN PARTITION

Partition of particles often depends strongly on the type of salt present in the system [1, pp. 118-121]. This steering effect increases with the electrical charge of the particle. The effect of the salt on the partition is therefore due to electrical forces. When the cation and anion of the salt differ

in their relative affinity for the phases a different distribution of the ions will occur at the interface, giving rise to a
potential difference between the phases, a so called interfacial
potential [4]. If the anion surpasses the cation in affinity for
the upper phase, as with NaCl, the upper-phase side of the interface will be negatively and the lower side positively charged.
Biological particles, which in general are negative, will distribute primarily into the lower phase, compared with the behaviour in a system with zero interfacial electrical potential.
If the salt creates an interfacial potential of the opposite
sign, i.e. upper-phase side positively and lower side negatively
charged, then the particles can be driven into the upper phase.
When the affinities of the particles for the two phases are
nearly equal, particles will accumulate at the interface. The
charge of the particles can also be estimated by comparing the
effects of two neutral salts on the partition. Studies done on
proteins can be used as a guide [4].

The composition of the phases of a given system can be found
using a phase diagram (Fig. 1). The curved line is called the
binodial curve and represents the minimum polymer concentrations
to get two phases; 0 is the critical point where the compositions
and the masses of the two phases become indistinguishable. A
phase system of composition T in the phase diagram will have an
upper-phase composition U and a lower-phase composition L. Pairs
of points such as U and L are called nodes, and lines joining
them are called tie-lines. The mass ratio and, therefore,
approximately the volume ratio between upper and lower phases
is equal to the ratio between the distances TL and UT. The
binodial curve and the tie-lines change with temperature and
depend on the molecular weights of the polymers [1]. There may
also be small variations between polymers of different batches.
Changing polymer concentrations is equivalent to moving from one
point in the phase diagram to another point. The further the
composition of the phases is removed from the critical point,
the more the accumulation of particles will be at the interface
(for illustrative data [5] with chloroplasts, see p. 89). Phase
systems with compositions close to the critical point are very
sensitive to small changes in temperature and polymer concentration.

The phase compositions will affect the interfacial potential
of the system [6]. This potential in a dextran-PEG system with
various salts is shown in Fig. 2, measured by partitioning of
proteins with known net charge [4]. By choosing systems closer
to the critical point (lower concentrations of the polymers) the
interfacial potential will decrease in magnitude, and vice versa.
Changing the mol. wt. of the polymers or the temperature will
also change the zero point of the interfacial potential scale.
For systems with PEG of mol. wt. 6000, sodium acetate gives an

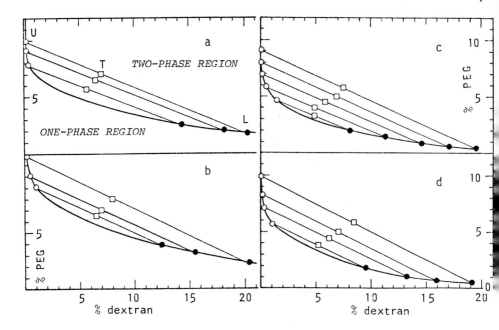

Fig. 1. Phase diagrams of the systems water-dextran-PEG (dextran
mol. wt. 500,000). The PEG was of mol. wt. 4,000 in a and b,
and 6,000 in c and d. Temperature: a, 0°; b, 20°; c, 3°; d, 20°.
□ , system composition; o, composition of upper phase; ●, composition of lower phase. Data from [1], by permission.

interfacial potential close to zero whereas, e.g., sodium chloride gives a 'negative' upper phase.

   Also the separation capacity of the biphasic system, i.e. how
much two kinds of particles differ in their partition, depends
on several factors, such as the mol. wt. and concentration of
the two polymers (PEG and dextran), the salts present and their
concentrations, pH and temperature [7, 8].

COUNTER-CURRENT DISTRIBUTION (CCD)

If the sample of material contains particles with specific properties, e.g. enzymatic activities, their distribution can be
studied even in impure mixtures. Through comparison with the
overall distribution of material the separation properties of
the system are directly determined. For better separation, or where
no specific method exists, a multi-stage procedure such as CCD
can be used [1], suitably with a special thin-layer apparatus
developed by Albertsson [1], or with a Craig-type apparatus (cf.

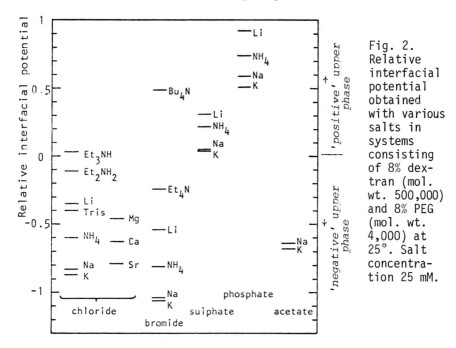

Fig. 2. Relative interfacial potential obtained with various salts in systems consisting of 8% dex- tran (mol. wt. 500,000) and 8% PEG (mol. wt. 4,000) at 25°. Salt concentra- tion 25 mM.

I.A. Sutherland's new continuous-flow method, *this vol.*). Then the distribution of total material can be determined by a non-specific method, e.g. turbidometrically. The resulting CCD-curve (for an example, *see* [3]) shows directly whether the material is divided into discrete fractions. Two points are notable.– (a) Even one-cell cultures can give rise to more than one fraction if the cells vary in their properties during the cell cycle and the cell culture is not synchronized [9, 10]; (b) the particles in a mixture may show a different distribution pattern from that of the pure particles when there are inter-action phenomena.

The separatory capacity can be measured by the difference in the log G values ($\Delta \log G$) [8], where G is the partition ratio given as the amount of material in the upper (mobile) phase divided by the amount in the stationary part of the system (interface and lower phase). By using the approximate relation [11]: $G = \hat{i}/(n-\hat{i})$ where $\hat{i}$ is the tube in the CCD-train with maximal amount of a certain component and n is the number of transfers, $\Delta \log G$ can be expressed by

$$\Delta \log G = \log [\hat{i}_A(n-\hat{i}_B)/[\hat{i}_B(n-\hat{i}_A)]$$

between two components (diagram peaks) A and B. By varying the volume ratio of the two phases, G, can be adjusted within some practical limits. Optimal separation by CCD is obtained when the product of the G-values (if the components are present in

comparable amounts) is close to unity [11]. Since the variation
of G by changing the phase volume ratio is limited by the CCD
apparatus used, G has to be adjusted by other means *(see below)*.

If no CCD apparatus is available, CCD can be carried out in a
row of test tubes. A multi-stage partition with 9 transfers is
here described. A large (50 g) phase system without sample is
prepared in a separatory funnel, and 1.5 ml of lower phase and
0.2 ml of upper phase are pipetted into each tube except no. 0
(Fig. 3). The contents of no. 0, i.e. sample in 1.5 ml lower
phase and 2 ml upper phase, are mixed by 20 inversions and
allowed to settle, then 1.8 ml of upper phase is transferred to
tube no. 1 and 1.8 ml fresh upper phase is added to tube no. 0.
Tubes 0 and 1 are again inverted 20 times and the phases allowed
to separate. The upper phases (1.8 ml) are transferred again one
step to the right and the fresh upper phase is added to the 0-
tube. Again 20 inversions and transfers of the upper phases are
made and the process is repeated until all the tubes contain both
phases. For analysis, water, buffer or sucrose solution (3.5 ml)
is added to each tube in order to break the phase systems.

PARTITION OF PARTICLES WITH CHARGED PEG

The partition of particles can be strongly influenced by intro-
ducing charged groups on a fraction of PEG [3, 10]. If PEG is
positively charged, e.g. trimethylamino-PEG (TMA-PEG), it will
attract the negative particles into the upper phase. If, however, the particles in an ordinary system are in the upper phase
they can be moved to the interface or the lower phase by using
PEG with negatively charged groups, e.g. PEG-sulphonate (S-PEG)
or carboxymethyl-PEG (CM-PEG). When systems with PEG of mol. wt.
4,000 and at low ionic strength (<5 mM) are used, little (<5%) of
the total PEG has to be replaced by its charged form, to extract
all the material from one phase to the other. Charged PEG of
mol. wt. 6,000 has considerably less effect on the partition,
and a 20-50 times larger concentration is needed to obtain the
same effect. Using charged PEG the concentration of other electrolytes in the system, e.g. buffer, must be kept low. By varying the percentage of charged PEG it is easy to find a system in
which the particles distribute equally between the upper phase
and the interface plus lower phase, i.e. G = 1. This partition
is well suited when the mixture is to be resolved with CCD. It
is also probable that the separatory capacity of the system, as
in the case of protein fractionation [8], is enhanced by the
charged PEG derivatives. This technique, more fully considered
previously [*this series:* 3], has been used for separation of
various membrane fractions from the electric organ of *Torpedo
marmorata* [12].

Highly charged dextran, e.g. commercial DEAE-dextran, has also
been used to drive the partition. Because of the very high degree

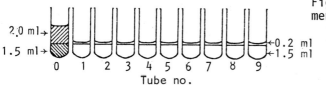

Fig. 3. Arrangement for manual CCD experiment with 9 transfers.

of substitution, these derivatives have lost their dextran character and can be found either in the upper or in the lower phase depending on the electrolytes present.

HYDROPHOBIC LIGANDS AND AFFINITY PARTITION

Specific ligands covalently attached to one of the polymers can selectively change the partition of membrane particles. Hydrophobic ligands or ligands with high specificity for membranes can be used. Fatty acids esterified with PEG have affinity for erythrocytes [13], varying with the fatty acids and with the erythrocyte species. A correlation between partition and lipid composition has been found. Recently liposomes have been used as a model system for the partition of membrane particles [14]. The polar head group of the phosphatides was shown to play a dominant role in determining the partition, while the cholesterol content and the degree of unsaturation of the phosphatides fatty acids are of less importance.

The chain-length of the fatty acid esterified with PEG has to exceed a certain minimum to allow extraction of biological particles into the upper phase. Since this critical length varies for different kinds of particles, a mixture of particles can be resolved by using certain PEG-esters: thus a mixture of intact and broken chloroplasts can be separated using PEG-caprinate [15]. By use of PEG-laurate, particles present in a liver homogenate can furnish a fraction enriched in nuclei and another in mitochondria.

Hydrophobic ligands can thus be used to detect differences in surface properties of membrane particles. In general, polymer-bound ligands offer a simple way to study the binding properties of the membrane surfaces. The technique with biospecific ligands has been used for the selective partition of membrane vesicles from the electric organ of *Torpedo californica* with the ligand *p*-trimethylammoniumphenylamine which binds to the membrane-bound acetylcholine receptors [16].

Preparation of affinity polymers

Carboxylic acids can be bound directly to PEG with the aid of carbodiimides [17] or by using the acid chloride [18] or anhydride. Dibasic carboxylic acids have also been esterified with PEG, forming semiesters, while the remaining acid group has been

used for binding molecules containing hydroxyl or amino groups.
PEG can also be carefully oxidized to carboxy-methyl-PEG with
$KMnO_4$ in neutral aqueous solution, and then reacted with alcohols
or amines [19]. By treating PEG with $SOBr_2$ in toluene, bromo-
PEG is obtained [3, 7], which can be further reacted, e.g. with
amines. By these methods a number of interesting biological substances, such as steroids, coenzymes and vitamins, have been
bound to PEG; but their effect on the partition properties has
been tested mainly on proteins [19].

STEPS TO OPTIMIZE THE PARTITION

How to get more material in the upper phase:
1. Choose another salt according to Fig. 2.
2. Choose a phase system closer to the critical point.
3. Use PEG with charge or with ligand that will attract the
material.
4. Choose another phase system with lower mol. wt. of PEG.

How to get more material in the lower phase:
1. Choose another salt according to Fig. 2.
2. Choose a phase system closer to the critical point.
3. Use PEG with charge of the same sign as the material or
dextran-bound ligands.
4. Choose another phase system with lower mol. wt. of dextran.

Fuller guidance on optimization with charged PEG is available
(*this series* [3]). The possibility of separating complex mixtures in one partition step has increased by the use of 3-phase
systems consisting of PEG, Ficoll^R and dextran in water [20].
The particles can in this system be located in 5 places: in the
3 phases and at the 2 interfaces between them.

APPLICATIONS

Phase systems have found many applications, e.g. for fractionation or study of erythrocytes [13, 21], mitochondria [22], chloroplasts [23, 24, & Fig. 4], plasma membranes [25], HeLa cells [26],
yeast cells [10], ribosomes [27], algae [9], bacteria [28, 29],
synaptic vesicles [12, 16, 30] and liposomes [14]. Thus, a CCD
separation of *Chlorella pyrenoidosa* into young and old cells has
been illustrated (*this series*: [3]; *also* [9]), and likewise erythrocytes [31]. Some general reviews cover partition in biphasic systems [1, 32, 33]. Further publications on separating cells are cited
at the end of the main reference list.

References
1. Albertsson, P.-Å. (1971) *Partition of Cell Particles and
   Macromolecules, 2nd edn.,* Almqvist & Wiksell, Stockholm.
2. Winlund Gray, C. & Chamberlin, M.J. (1971) *Anal. Biochem. 41,*
   83-104.

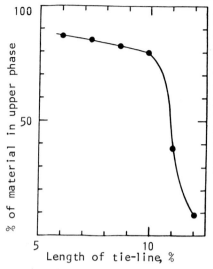

Fig. 4. Partition of spinach chloroplasts as affected by polymer concentrations. The % of total material collected in the upper phase is given as a function of length of tie-line, measured in the polymer concentration scale. System: 6-7% dextran (mol. wt. 500,000), 6-7% PEG (mol. wt. 4,000), 5 mM sodium phosphate buffer pH 7.8 at 2°. *Data from [5].*

3. Johansson, G. (1973) in *Preparative Techniques (Methodological Developments in Biochemistry, Vol. 2)* (Reid, E., ed.), Longman, London, pp. 155-162.
4. Johansson, G. (1974) *Acta Chem. Scand. B28,* 873-882.
5. Karlstam, B. & Albertsson, P.-Å. (1972) *Biochim. Biophys. Acta 255,* 539-552.
6. Johansson, G. (1978) *J. Chromat. 150,* 63-71.
7. Johansson, G., Hartman, A. & Albertsson, P.-Å. (1973) *Eur. J. Biochem. 33,* 379-386.
8. Johansson, G. & Hartman, A. (1974) *Proc. Internat. Solvent Extract. Conf., Lyon,* pp. 927-942.
9. Walter, H., Eriksson, G., Taube, Ö. & Albertsson, P.-Å. (1970) *Exp. Cell Res. 64,* 486-490.
10. Johansson, G. (1974) *Mol. Cell. Biochem. 4,* 169-180.
11. Craig, L.C. (1962) in *Comprehensive Biochemistry, Vol. 4* (Florkin, M. & Stotz, E.H., eds.), Elsevier, Amsterdam, pp. 1-31.
12. Hartman, A. & Heilbronn, E. (1978) *Biochim. Biophys. Acta 513,* 382-394.
13. Eriksson, E., Albertsson, P.Å. & Johansson, G. (1976) *Mol. Cell Biochem. 10,* 123-128.
14. Eriksson, E. & Albertsson, P.Å. (1978) *Biochim. Biophys. Acta 507,* 425-432.
15. Johansson, G. & Westrin, H. (1978) *Plant Sci. Lett. 13,* 201-212.
16. Flanagan, S.D., Barondes, S.H. & Taylor, P. (1975) *J. Biol. Chem. 251,* 858-865.
17. Johansson, G. (1976) *Biochim. Biophys. Acta 451,* 517-529.
18. Shanbhag, V.P. & Johansson, G. (1974) *Biochem. Biophys. Res. Commun. 61,* 1141-1146.
19. Johansson, G., *to be published*.
20. Hartman, A. (1976) *Acta Chem. Scand. B30,* 585-594.

21. Walter, H. (1977) in *Methods of Cell Separation*, Vol. 1 (Catsimpoolas, N., ed.), Plenum, New York, pp. 307-354.
22. Ericsson, I. (1974) *Thesis*, University of Umeå, Sweden.
23. Larsson, C., Collin, C. & Albertsson, P.Å. (1971) *Biochim. Biophys. Acta 245*, 425-438.
24. Larsson, C. & Andersson, B. (1979) *Plant Organelles* (Vol. 9), *this series*, pp. 35-46.
25. Lesko, L., Doulon, M., Marinetti, G.V. & Hare, J.D. (1973) *Biochim. Biophys. Acta 311*, 173-179.
26. Pinaev, G., Hoorn, B. & Albertsson, P.Å. (1976) *Exp. Cell Res. 98*, 127-135.
27. Pestka, S., Weiss, D. & Vince, R. (1976) *Anal. Biochem. 71*, 137-142.
28. Magnusson, K.E. & Johansson, G. (1977) *FEMS Microbiol. Lett. 2*, 225-228.
29. Stendahl, O., Magnusson, K.E., Tagesson, C., Cunningham, R. & Edebo, L. (1973) *Infect. Immun. 7*, 573-577.
30. Flanagan, S.D., Taylor, P. & Barondes, S.H. (1975) *Nature (Lond.) 254*, 441-443.
31. Walter, H. & Selby, F.W. (1966) *Biochim. Biophys. Acta 112*, 146-153.
32. Albertsson, P.Å. (1970) *Adv. Protein Chem. 24*, 338-341.
33. Albertsson, P.Å. (1978) *J. Chromat. 159*, 111-122.

SUPPLEMENTARY REFERENCES concerned with cell separation. -

TISSUE CELLS:

Brunette, D.M., McCulloch, E.A. & Till, J.E. (1968) *Cell Tissue Kinet. 1*, 319-327. - *Fractionation of Suspensions of Mouse Spleen Cells by Counter Current Distribution.*
Walter, H., Krob, E.J., Ascher, G.S. & Seaman, G.V.F. (1973) *Exp. Cell Res. 82*, 15-26. - *Partition of Rat Liver Cells in Aqueous Dextran-Polyethylene Glycol Phase Systems.*
Walter, H. & Krob, E.J. (1975) *Exp. Cell Res. 91*, 6-14. - *Alterations in Membrane Surface Properties During Cell Differentiation as Measured by Partition in Aqueous Two-Polymer Phase Systems: Rat Intestinal Epithelial Cells.*
Malmström, P., Nelson, K., Jönsson, Å., Sjögren, H.O., Walter, H. & Albertsson, P.-Å. (1978) *Cell Immunol. 37*, 409-431. - *Separation of Rat Leukocytes by Countercurrent Distribution in Aqueous Two-Phase Systems.*

TISSUE-CULTURED CELLS:

Gersten, D.M. & Bosmann, H.B. (1974) *Exp. Cell Res. (I) 87*, 73-78; *(II) 88*, 225-230. - *Behaviour in Two-Phase Aqueous Polymer Systems of L5178Y Mouse Leukemic Cells. I: In the Stationary Phase of Growth; II: The Lag and Exponential Phases of Growth.*
Kessel, D. (1976) *Biochem. Pharmacol. 25*, 483-485. - *Alteration of Cell [murine leukaemia] Permeability Barriers by Amphoterium β-Deoxycholate (Fungizone) in Vitro. See also ref. [26].*

#B-6

FREE-FLOW ELECTROPHORESIS FOR THE ISOLATION OF HOMOGENEOUS POPULATIONS OF BIO-PARTICLES, PARTICULARLY OF CELLS

HANS-G. HEIDRICH and KURT HANNIG,
Max-Planck-Institut für Biochemie,
Martinsried b. München, W. Germany.

Free-flow electrophoresis enables soluble material (proteins, nucleic acids) and bio-particles (cells, organelles, membrane systems) to be isolated in quantity, under crucial conditions. Relevant parameters include ζ-potential, band-broadening effects (velocity profile, electro-osmosis), ionic strength of separation buffers (separation according to charge or size), coating of the separation chamber walls, and conditions during separation.

The techniques for preparing the biological material to be separated are also vital. Comments are made on cell-isolation techniques, homogenization methods and aggregation, and isolation media. Consideration is given to the separation of B and T lymphocytes, of kidney cortical cell populations (rabbit), and of malaria parasites and of erythrocytes containing or lacking these [cf. other contributions to this vol.*– Ed.]

Many improvements have recently been made to the technique of free-flow electrophoresis [1] resulting in better and more distinct separations and resolutions[†]. These improvements have resulted from theoretical consideration of the fundamental parameters of the technique as well as from changes in conditions, as will now be discussed.

THEORETICAL CONSIDERATIONS

1. The electric double layer

The charge density (i.e. the real charge) on the membrane of a bio-particle (cell, organelle, membrane vesicle) cannot be estimated from the electrophoretic mobility of the particle

---

*resp. M.J. Owen & M.J. Crumpton; R.G. Price; S.M. Lanham.

[†]See also, for lysosomes, R. Stahn et al. in *Separations with Zonal Rotors* (ed. E. Reid, this series).

alone. Several parameters which are influenced by the buffer milieu have also to be taken into consideration. A bio-particle in an electrolyte of pH 7.2-7.4 and in an electric field is normally negatively charged and is surrounded by an electric double layer (Fig. 1). This double layer has two regions. The inner region (Stern layer) is the hydrated layer which surrounds any particle in an aqueous medium and into which, because of electrostatic interactions, mainly cations from the medium enter. These ions alter the real negative surface charge of the bio-particle, and the resulting potential at the surface of the Stern layer (surface of shear) is the electrokinetic or the zeta-potential of the particle. The zeta-potential is the apparent particle charge and depends on the ion concentration in the electrolyte. The real surface charge can be approximated by an extrapolation of the apparent particle charge to an in-finitely dilute solution; but for the interpretation of electro-phoretic separations this real charge is of no practical impor-tance. The measurable characteristic charge of the particle corresponds to the zeta-potential.

The outer region of the electric double layer of the bio-particle is a diffuse cloud of ions (Debye-Hückel layer, or Gouy-Chapman layer), distributed around the particle according to the influence of electrical forces and random thermal motion. The ion cloud around the particle is influenced by the dielectric constant of the electrolyte and by the electric field. The latter causes a deformation of the cloud in the direction opposite to that in which the particles move (relaxation effect).

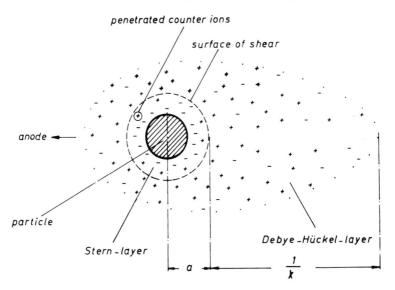

Fig. 1. The diffuse electric double layer.

2. Separation according to charge or to size

The relaxation effect opens up one of the crucial questions in
free-flow electrophoresis: does one separate according to charge
or according to particle size ? The characteristic description
of the ion cloud is given by $\kappa \cdot a$, i.e. the ratio of the particle
radius (a) to the extension of the ion cloud ($1/\kappa$). In the
electric field the ion cloud will be deformed and a dipole
moment results. At very large or at very small extensions of
the cloud the effect is negligible. But calculations by
Wiersema [2], proven experimentally, indicate that in the range
$1 < \kappa \cdot a < 100$ the particles are electrophoretically deflected
not only according to the zeta-potential but also according to
size. Such conditions can be created by variation either of the
particle size (which is impossible in practice since the size of
the particles to be separated is fixed) or of the buffer milieu
(which influences κ).

Fig. 2 shows that at a low buffer conductivity latex beads
can be separated according to size because $\kappa \cdot a$ is between 1 and
100 [3]. However, when the conductivity of the buffer is increased, and $\kappa \cdot a$ is larger than 100, the particles are deflected
only according to their zeta-potential, and their size does not
play any role in the electrophoretic deflection. For separation
of bio-particles according to charge only, the conductivity of
the separation medium should be between 500 and 600 $\times$ $10^{-6}\Omega\text{cm}^{-1}$
(5-6 $\times$ 10^2 µmhos), since bio-particles normally have diameters
in the range 100 nm to 50-80 µm. Preliminary experiments in our
laboratory have shown that erythrocyte vesicles can be separated
from erythrocyte ghosts electrophoretically on the basis of
their size, using low ionic strength buffers. However, the
question of membrane integrity arises when using such buffer
systems.

3. Band-broadening

There are two main effects in free-flow electrophoresis which
cause band-broadening and a clear loss of resolution [4, 5].
These are the influence of the velocity profile, and that of the
electro-osmotic flow profile, on the particles close to the
chamber walls (Fig. 3). The influences act in opposite directions and to a certain extent they compensate each other. The
conditions for complete compensation, however, can be fairly
easily calculated from the theory.

The horizontal displacement (s_e) of the particles can be
described by the equation

$$s_e = \frac{2}{3} E\tau_m \cdot u_e \left(1 - \frac{1}{2} \frac{u_w}{u_e}\right) + \frac{2}{3} E\tau_m \cdot u_e \cdot S \cdot F_o \qquad (1)$$

where E is the strength of the electric field applied, τ_m the

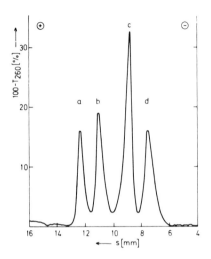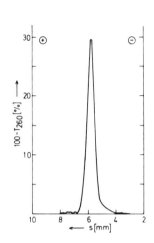

Fig. 2. Electrophoretic separation of latex particles according
to size (left) or to charge (right).
In the run presented in the left part latex particles (diameter:
a = 2.02, b = 1.01, c = 0.48, d = 0.234 μm) were separated in a
Tris/borate buffer 3×10^{-4} M, pH 8.6, spec. conductivity 12.6 ×
$10^{-6}\Omega^{-1}cm^{-1}$; $\kappa = 4.1 \times 10^5 \times cm^{-1}$. Therefore $\kappa \cdot a$ has values
between 41.4 - 4.8, and the spearation is according to size. In
the run presented in the right part the same particles were
separated in a Tris/borate buffer 0.3 M, pH 8.7, spec. conducti-
vity $1,100 \times 10^{-6}\Omega^{-1}cm^{-1}$. Here $\kappa \cdot a > 200$ and the separation
is according to charge.

---

mean transit time through the field, u_e the electrophoretic
mobility of a particle, u_w the electro-osmotic mobility. S des-
cribes the parabolic flow curve in a particular separation chamber. F_o is the expression for the flow at the chamber walls
and is expressed by $(1 + u_w/u_e)$, since disturbances derived from
temperature and electric field can be neglected. If u_w/u_e is
substituted by the zeta-potentials, one gets $u_w/u_e = -\zeta_w/\zeta_e$.
The sign in this term is negative as long as the flow of the
liquid along the chamber walls is opposite to the movement of
the migrating particle. After substitution $F_o = (1 - \zeta_w/\zeta_e)$,
and if $\zeta_w = \zeta_e$, F_o becomes zero. In this case the second expression of equation (1), which stands for processes close to
the chamber wall, also becomes zero and equation (1) is reduced
to $s_e = \frac{1}{3} E T_m u_e$. Thus the two effects on the chamber walls
can be compensated by adjusting the zeta-potential of the chamber walls to the zeta-potential of the particles to be separated. For routine separations the effects can be sufficiently

Fig. 3. Relationship
between flow profiles
and band-broadening.

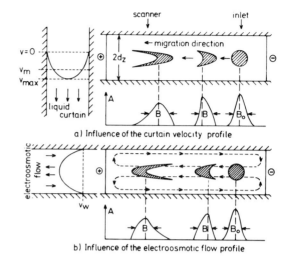

well compensated by
coating the chamber
walls with a physio-
logical protein.
Albumin has been used
successfully. After
albumin coating the
glass walls have a
zeta-potential of
-34 mA, comparable
to the zeta-poten-
tials of cells,
organelles and mem-
brane vesicles which
are in the range of
-27 to -45 mA in the normal buffers used. It should be pointed
out that all kinds of proteins can be coupled to glass *via*
suitable silane-derivatives. Thus any particular type of important bio-particle can be separated optimally by having the same
zeta-potential on the chamber walls as the bio-particle possesses.

PRACTICAL APPLICATIONS

1. Separation conditions

From all these considerations it follows that a good buffer for
free-flow electrophoresis experiments with bio-particles should
have a pH of 7.2-7.4 and a conductivity in the range of 5-10 ×
10^2 μmhos (better higher if the cooling system of the apparatus
used permits very high currents). The osmolarity should be
matched to the osmotic properties of the particles to be separated.

Triethanolamine/acetic acid and HEPES have been shown to be
excellent buffers for the separation of bio-particles. However,
the latter is generally unsuitable for routine work because of
the large amounts needed for the electrode buffer and the consequent high cost. Triethanolamine (free base !) must be colour-
less and of the highest purity, and the buffers have to be
freshly prepared. Other salts, such as Ca^{2+}, Mg^{2+}, Mn^{2+} or EDTA,
can be added to the buffers in μmolar to mmolar concentrations.
For adjusting the osmolarity, sucrose, glucose, ribose, sorbitol
or glycine can be used. Glycine has the advantage of having no
influence on the viscosity of the buffer, but has proved unsuitable for kidney cells in our experiments.

The coating of the separation chamber is carried out with a 3%

albumin solution in the separation medium for 60 min, followed
by an extensive wash with water. The coat has to be renewed
prior to each experiment. Other conditions in the chamber are
also of importance for a good resolution. An efficient and uniform cooling system should be used in order to avoid thermoconvec-
tions and to guarantee a high electric field. If the field is
too weak it will not be possible to have a sufficiently fast
buffer flow through the chamber. Generally the buffer flow
should not be so slow that the sample is deflected into the
furthest fractions, since a wide deflection decreases resolution
by increasing band width. The resolution is also increased by
using a very fine fractionation device (many chamber outlets per
cm).

2. Preparation of the material to be separated

The condition of the material to be separated is also of importance for an optimal separation. When separating cells, the
fundamental pre-requisite is a suspension of intact single cells
without aggregates or cell clumps. The cell surface structures
must be as unaltered as possible. Cell suspensions are normally
made using weak mechanical forces (e.g. from bone marrow, spleen,
thymus, lymph nodes) or enzymes such as collagenase, hyaluronidase, trypsin or dispase (e.g. from liver, fat tissue, kidney,
carcinomas). Collagenase can alter the cell surface by splitting
off receptors [6, *inter alia*]. Dispase, or a combination of
dispase and other enzymes, used for the disintegration of rat
kidney cortex alters the morphology of proximal tubule cells insofar as all the microvilli are lost [7]. Trypsin alone, however, appears not to destroy kidney cells [8]. In our own
experiments enzymes are avoided as much as possible. For example
we were not able to achieve a successful separation of kidney
cortex cells from rabbit after using enzymes for tissue disintegration; but citrate perfusion of the kidney resulted in a
suspension of intact cells possessing the microvilli and in a
successful electrophoretic separation [9]. There are no general
rules for producing good cell suspensions, and appropriate procedures must be developed for each cell type used.

This is also true for homogenization methods used in isolating
organelles and membranes from tissue. The often observed
aggregation of homogenates during electrophoretic runs results
from over-harsh homogenization of the tissue and subsequent
damage to the cell nuclei. The DNA and chromatin which are
thereby released are adsorbed onto the membranes and cannot be
washed off. As a consequence clumping is accordingly observed
in the electric field and precludes successful separation.
Aggregates can be partially removed by centrifugation (100 g for
30 min) or by treating the homogenate with a DNAse solution (20-
30 μg/ml) at room temperature for 10 min. However, aggregation
can be avoided completely if the first homogenization step is

carried out very carefully in a 1 : 10 dilution, in buffer, of the tissue after mincing or passage through a tissue-press, particularly if the homogenate is centrifuged at 100 g for 10 min prior to electrophoresis.

2. Immune competent cells: B and T lymphocytes

In the immune system it is of interest to investigate the correlations between the differentiating cells and their environment, and also to ascertain at which level of differentiation the immune competence of the cell is formed and under which conditions self-tolerance is found. Free-flow electrophoresis allows the separation of immune competent peripheral lymphocytes into a population of thymus-dependent T cells and a population of thymus-independent B cells [10]. Based on these results five sub-populations of T cells [11] have been isolated from thymus using a combination of free-flow electrophoresis and 1 g sedimentation. These populations were highly enriched. They represent stages in a differentiation sequence which is characterized by increasing maturity. The isolated cell populations show functional differences which allow of their characterization and a logical classification as well as a selective inhibition of their functions. Sub-populations have also been isolated from B cells [12].

3. Cell populations from rabbit kidney cortex

Different differentiation levels or functional states of bioparticles are usually expressed by different surface structures, i.e. by different surface charges. Kidney cells have specific functions and obvious morphological differences. Therefore one could expect that the different cell types would behave differently in an electric field. Homogeneous populations of single and viable kidney cells have not been described unequivocally; they could be especially useful for diagnostic-therapeutic as well as biological-biochemical studies, e.g. relating to renin or kallikrein. Cell populations have been isolated as follows [9].

Rabbit kidneys were perfused with 100 ml of Earle's medium, pH 7.4, containing no enzymes but 3 mM sodium citrate. Then the cortices were removed and a single-cell suspension was made. The medium contained 11 mM triethanolamine, 11 mM acetic acid, 2.5 µM $MgCl_2$, 5 µM $CaCl_2$, 5 mM glucose, 285 mM sucrose, and 5 mg BSA per 1,000 ml; KOH to give pH 7.5 (osmolarity 355 m-osM; conductivity of 6.4×10^2 µmhos). The suspension was centrifuged three times for 5 min at 150 g to remove debris and cell organelles and damaged cells, filtered and then subjected to free-flow electrophoresis which was carried out in the described medium at 5° at 230 mA, 132 V/cm; buffer flow 2 ml/fraction/h; sample injection 40×10^6 cells/ml/30 min.

After cell counting a main peak with shoulders on the anodic and cathodic sides was found (Fig. 4). From thin sections of the

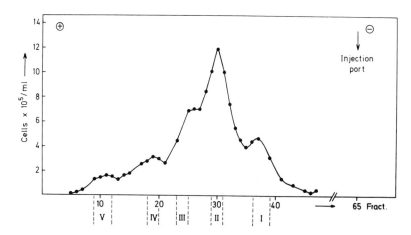

Fig. 4. Cell count profile from a run of a kidney cortex cell
suspension in an FFV apparatus. All the cells were deflected
towards the anode. Roman numerals indicate the different pools
(see text).

---

separated cells it was impossible to draw any conclusions concerning the morphology of the cells in the main fractions and the
shoulders because of uncertainty about the plane of sectioning
(Plate 1). However, morphological differences amongst the cells
in the electrophoretic fractions could be clearly determined by
scanning microscopy: most of the cells in fractions 35-40 were
erythrocytes; the cells in the main peak (II) were proximal
tubule cells with long microvilli (Plate 2), while the anodic
shoulder (III) contained cells with short and fewer microvilli.
The cells in fractions 18-22 were renin-positive and were much
smaller than the proximal tubule cells. Biochemical characterization is being pursued, as is checking of the vitality of the
cells which in culture incorporate uridine. Similar experiments
have been tried with rat kidney cortex. Our own experiments are
still unsatisfactory since the cells from rat kidney appear to be
less stable than those from rabbit. Thus other isolation procedures and culture conditions have to be used. With rat kidney,
however, proximal tubule cells have been isolated by Kreisberg
et al. [8] using free-flow electrophoresis.

4. Separation of malaria-infected and uninfected erythrocytes, and isolation of the erythrocytic stages of malaria parasites

(Plasmodium vinckei)

In studying malaria fundamental difficulties arise, since the
parasite grows inside the host cell (erythrocyte) and overlaps
metabolically with it. Metabolic and other studies, e.g. on cellular organization or immunochemistry, would be simplified if intact parasites were available in preparative quantities. As now

described, free plasmodia can be rapidly isolated from their host cells or infected from uninfected erythrocytes [13].

Mice were infected with 10^8 erythrocytes from a donor mouse with a parasitaemia of 30-40%. After 3-4 days blood was harvested and the erythrocytes were washed in electrophoresis medium consisting of 10 mM triethanolamine, 10 mM acetic acid, 0.1 mM MgSO₄, 0.25 M sucrose; NaOH to give pH 7.4 (osmolarity 315 mosM;

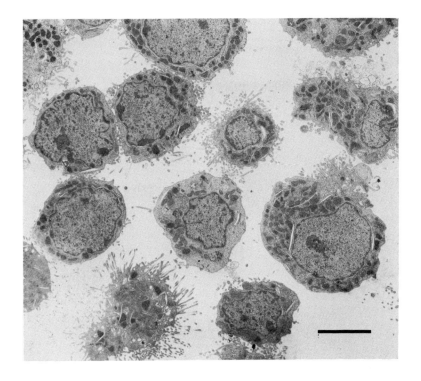

Plate 1.   Thin section through cells from an electrophoresis run.   All the cell-containing fractions were pooled for preparing this specimen, in order to show the morphological condition of the cells after the run.   Bar is 5 μm.

conductivity 5.3×10^2 μmho). For erythrocyte separation these cells were injected into an FFV Free-Flow Electrophoresis apparatus (Bender & Hobein, München) at 5°, 180 mA and 135 V/cm. For isolating free plasmodia the erythrocytes were treated with a concanavalin A solution for 10 min in ice and then ejected three times through a hypodermic needle. The resulting suspension was washed three times in electrophoresis medium by centrifugation at 2,000 g for 10 min at 4°. This material was injected as for erythrocytes.

When erythrocytes from infected mice were run electrophoretically, two bands were seen. The cells from the more anionic moving

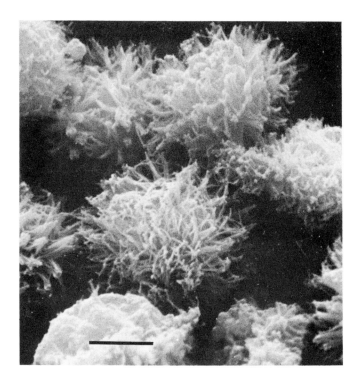

Plate 2. Scanning microscopy on cells from pool II from the run in Fig. 4.
All the cells possess long microvilli and are proximal tubule cells. Bar is 5 μm.

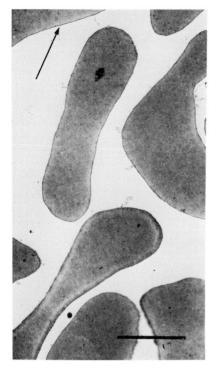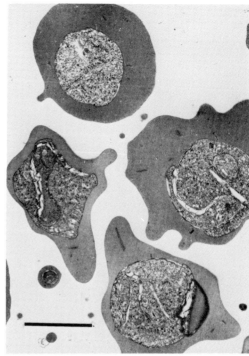

Plate 3. *Left portion.* — Erythrocytes from an infected mouse, after electrophoresis, possessing the same electrophoretic mobility as erythrocytes from an uninfected animal. *They have a more anionic nature than parasitized erythrocytes. Their shape is still biconcave. Seldom can the start of a parasitaemia be seen (arrows).* Bar is 2 µm.
Right portion. — Erythrocytes from an infected mouse after electrophoresis. *These erythrocytes have a weaker anionic nature than those on the left. They are all crenated and contain parasites in all kinds of developmental stages.* Bar is 2 µm.

band can be recognized morphologically as non-infected erythrocytes (Plate 3), and have the same electrophoretic mobility as erythrocytes from uninfected control animals. The cells from the less anionic moving band are infected erythrocytes (Plate 3). In this fraction all the intra-erythrocytic stages of plasmodia could be found with cytostomal cavities, food vacuoles, pigment, multiple nuclei, etc. Uninfected and infected erythrocytes evidently differ slightly in electric charge, and after injection the membrane presumably is altered chemically and/or physically, as expressed also by the difference in cell shape.

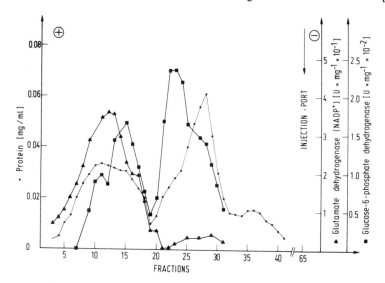

Fig. 5. Protein and enzyme profiles from an electrophoresis run
in which fractions 5-18 contain free parasites, fractions 20-24
'uninfected' erythrocytes (Plate 3, *left portion*), fractions 25-
31 infected erythrocytes (Plate 3, *right portion*), and fractions
32-40 unidentified material. Protein (●), glutamate dehydrogenase (NADP⁺) (▲) and glucose-6-phosphatase dehydrogenase (■)
were assayed.
Activities are expressed per mg protein; 1 unit = 1 μmol/min.

When the mixture of free plasmodia, infected and non-infected
erythrocytes and erythrocyte membranes was injected into the
electrophoresis apparatus, the protein distribution profile
shown in Fig. 5 was obtained. In the fractions of the broad peak
close to the anode there are free parasites in all developmental
stages, as also seen within the infected erythrocytes. This
fraction is absolutely homogeneous (Plate 4). The peak less
close to the anode contains in its left region erythrocytes
without parasites, in its right region erythrocytes with parasites (not shown here). The success of this separation could be
recognized not only from morphology but also from the distribution of marker enzymes tested. Glucose-6-phosphate dehydrogenase
(NADP⁺) can serve as a marker for erythrocyte stroma, and glutamate dehydrogenase which is not present in erythrocytes is a
marker for parasites. Clear distributions of the two enzymes can
be recognized (Fig. 5). The latter was found in all fractions
containing parasites or infected erythrocytes, with a notably
high specific activity corresponding to homogeneous parasites.
Glucose-6-phosphate dehydrogenase was present in parasites
(because of the food vacuoles) and in the erythrocyte fractions.
This clear separation between free plasmodia and infected and

Plate 4. Free parasites *(Plasmodium vinckei)* from fraction 10 of the electrophoretic run shown in Fig. 5. Bar is 5 μm.

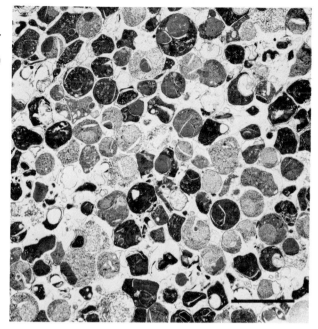

non-infected erythrocytes opens up new possibilities for studying malaria at the cellular and molecular level.

References

1. Hannig, K. (1972) in *Techniques of Biochemical and Biophysical Morphology*, Vol. 1 (Glick, D. & Rosenbaum, R., eds.), Wiley, New York, pp. 191-198; *see also* Hannig, K. & Heidrich, H.-G. (1974) *Methods in Enzmology 31*, 191-198.
2. Wiersema, P.H., Loeb, A.L. & Overbeek, J.Th.O. (1966) *J. Colloid Interface Sci. 22*, 78-84.
3. Hannig, K., Wirth, H., Meyer, B. & Zeiller, K. (1975) *Hoppe-Seyl. Z. Physiol. Chem. 356*, 1209-1223.
4. Kolin, A. (1967) *J. Chromatog. 26*, 180-193.
5. Strickler, A. & Sacks, T. (1973) *Prep. Biochem. 3*, 269-277.
6. Kawai, Y. & Spiro, R.G. (1977) *J. Biol. Chem. 252*, 6229-6244.
7. Köpfer-Hobelsberger, B. & Heidrich, H.-G. *In preparation.*
8. Kreisberg, J.I., Sachs, G., Pretlow II, T.G. & McGuire, R.A. (1977) *J. Cell Physiol. 93*, 169-172.
9. Heidrich, H.-G. & Dew, M.E. (1977) *J. Cell Biol. 74*, 780-788.
10. Zeiller, K. & Hannig, K. (1971) *Hoppe-Seyl. Z. Physiol. Chem. 352*, 1162-1167.
11. Zeiller, K., Holzberg, E., Pascher, G. & Hannig, K. (1972) *Hoppe-Seyl. Z. Physiol. Chem. 353*, 105-110.
12. Zeiller, K., Pascher, G. & Hannig, K. (1976) *Immunology 31*, 863-880.
13. Heidrich, H.-G., Rüssmann, L., Bayer, B. & Jung, A. (1979) *Z. Parasitenkd., in press.*

#C Populations from Tissues other than Liver

#C-1
ISOLATION OF KIDNEY GLOMERULI, TUBULAR FRAGMENTS AND CELL POPULATIONS

ROBERT G. PRICE
Department of Biochemistry,
Queen Elizabeth College,
Campden Hill,
London W8, U.K.

The kidney nephron is divided into highly specialized regions, each with its own characteristic morphology and function. For glomeruli and tubules, showing cell-type heterogeneity that reflects functional diversity, methods of isolation have been developed. Manual dissection of the kidney into cortex and medulla reduces the number of regions of the nephron present, and the tissue can be broken down into its major components by gentle homogenization, passing through a graded sieve or treatment with disaggregating agents. Separation of glomeruli and different tubular fragments can be achieved by differential sieving or gradient centrifugation. Cells suitable for subcellular, metabolic or tissue-culture studies may then be isolated.—

Source   (1) Glomeruli. (2) Nephron (known region ?) fragments.

Dissociation & EDTA and (1) enzymes, then pipetting; (2) ionic medium.
separation   For (2): gradient centrifugation or free electrophoresis.

Product   (1) Epithelial/endothelial/mesangial. (2) Tubular cell types.

If kidney cortex is taken, the cells recovered will be derived from the glomeruli, proximal and distal convoluted tubules as well as the blood vessels (Fig. 1). Studies involving homogenates or cells isolated from whole kidney are therefore of limited value. More meaningful data can be obtained by isolating the major regions of the nephron present in each zone of the kidney, and then isolating cell types from the fragments.

The requisite methods are of necessity empirical, and the experimental approach adopted will depend on the ultimate objective and the species of animal used. Young or neonatal tissue is readily broken down into glomeruli and tubular fragments. The presence of greater amounts of connective tissue in adult kidney hinders disaggregation. Gentle homogenization procedures are essential since more drastic procedures result in cellular breakdown. Renal tubular fragments are prone to aggregation, and preparations should be maintained in a non-humid environment at

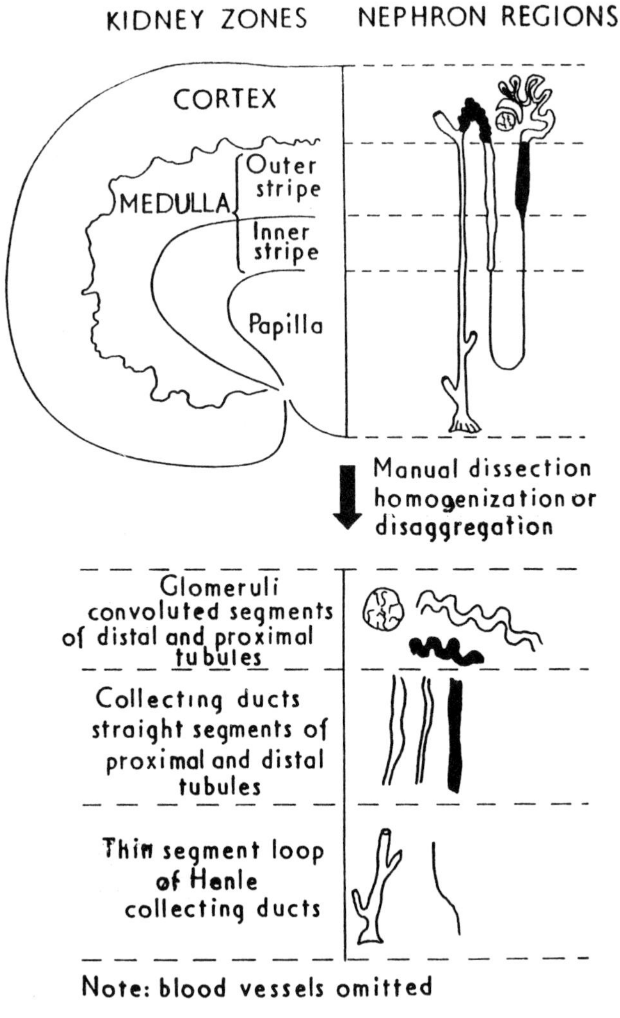

Fig. 1. Morpho-
logical inter-
relationship bet-
ween the regions
of the nephron
and zones of the
kidney.

KIDNEY ZONES　　NEPHRON REGIONS

CORTEX

MEDULLA　Outer stripe

Inner stripe

Papilla

Manual dissection
homogenization or
disaggregation

Glomeruli
convoluted segments
of distal and proximal
tubules

Collecting ducts
straight segments of
proximal and distal
tubules

Thin segment loop
of Henle
collecting ducts

Note: blood vessels omitted

4°. Siliconized
glassware or
plastic containers
should be used to
prevent the ad-
hesion of glome-
ruli to the sides
of the containers.
Preparations are
normally carried
out in physiolo-
gical saline or
a suitable tissue
culture medium.
Since each region
of the nephron
has a distinc-
tive morphology
(Fig. 1), we
normally monitor
each stage of a
particular iso-
lation by phase
microscopy.

**ISOLATION OF
REGIONS OF THE
NEPHRON**

Manual dissection

Whole nephrons can be manually dissected from rabbit or human
kidneys following maceration and treatment with HCl [1]. This
procedure has been used to follow changes in the architecture of
the developing nephron in embryonic kidneys [2] and in the detection of pathological lesions such as obstruction [1]. Treatment
with HCl is harsh, and this technique should be used with caution
for biochemical or physiological studies involving membrane
phenomena.

Microdissection

Elegant techniques involving the microdissection, with a sterop-
tic, of sections of the nephron from tissue frozen rapidly in

liquid nitrogen have been described [3]. Each section, after
identification by morphological examination of the adjacent sections, is freeze-dried and its enzymic properties determined by
fluorimetric or enzyme re-cycling procedures. These studies have
allowed the precise determination of the metabolic activity of
each region of the nephron [4], and may lead to convenient marker
enzymes for each region of the nephron.

SEPARATION OF ZONES OF THE KIDNEY

The kidney is an erectile organ, and loss of its blood supply results in the immediate collapse of the tubular lumen. Prior perfusion with saline or tissue culture medium [5] until the tissue
is blanched prevents some of the more serious effects of ischae-
mia and also removes blood cells which contaminate renal cell
preparations. When the rat is used as the experimental animal,
kidneys can be sliced longitudinally and the cortex removed with
curved scissors. Separation of the papilla from the medulla is a
simple matter in this species.

ISOLATION OF GLOMERULI AND GLOMERULAR CELLS

Rat kidney cortices can be conveniently disrupted by passing
through sieves [6]. A wide range of sieves are available and the
dimensions can be varied, depending on the size of nephron component to be isolated. Cortices can be stored in aluminium foil
after freezing to -40°, or processed immediately after dissection.
Thinly sliced cortices are buttered through a 170-mesh sieve in
0.15 M saline, and the resulting mixture is filtered through an
80-mesh sieve to remove tissue fragments. Finally, passage
through a 270-mesh sieve furnishes the glomeruli, accompanied by
a few contaminating tubules. Glomeruli thus prepared have been
used to prepare glomerular basement membrane [6] and for biosynthetic studies [7-9]. Glomeruli can be separated from tubules
present in disrupted cortical homogenates by density-gradient
centrifugation at low g-values on discontinuous Ficoll gradients
[7, 8]. Recently mechanical homogenization has been used to process large quantities of bovine kidney [10]; although this procedure is suitable for isolation of glomerular basement membrane,
milder techniques should be used for metabolic studies.

Glomeruli have been cultured [11] and the resultant cell monolayers are epithelial or fibroblast-like depending upon their
origin from either encapsulated or decapsulated glomeruli, respectively. Glomerular cells can be separated from isolated glomeruli
by sequential treatment with collagenase, EDTA and trypsin-collage-
nase-DNAase mixture followed by mechanical disruption by repeated
pipetting [12]. The cells were centrifuged (50 g, 5 min) with no
attempt to separate the different cell types present. Achievement
of such separation at this stage by centrifugation or free-flow
electrophoresis could avoid the cross-contamination which was
found [12] when the cells were cultured.

The different types of glomerular cells thus obtained retain
a distinctive morphological appearance, and epithelial and endothelial cells are readily distinguishable from the phagocytic
mesangial cells. Culture of isolated glomerular cells results
in a preponderance of phagocytic cells which can develop F_c and
C_3 receptors.

PREPARATION OF TUBULAR REGIONS AND CELLS

Disruption with sieves

The separation of glomeruli and tubular fragments from the renal
cortex depends on the difference in size and shape. Glomeruli
can be recovered by sieving as described above, and provided that
aggregation is avoided the remaining tubules can be separated into
proximal (80%) and distal (20%) tubular fractions by differential
sieving. The yield of distal tubules is normally low. Preliminary data from this laboratory suggest that the density of the
proximal convoluted tubule can be increased by protein loading
the rats before isolation, and such a procedure should allow the
separation of the two regions on a density gradient. Tubular
fragments aggregate in density-gradient media such as Ficoll [7].

The three major components of the nephron present in the
medulla (Fig. 1) can also be separated by sieving; but the
dimensions chosen for the sieves may vary since the length of the
fragments present is critical. The different morphology of the
straight segments of the proximal and distal tubules and the
collecting ducts can be distinguished by light microscopy (Fig. 1),
and the purity of each fraction can be assessed by this technique.
Separation of thin segments and collecting ducts from the rat
papilla has so far proved to be an intractable problem.

Disaggregating agents

Tubular fragments can be isolated following the break-up of
renal tissue with disaggregating agents. We have found 0.05%
collagenase [13] to be particularly useful; it should be injected
into renal pelvis prior to dissecting the cortex. Exposure of
the tissue to disaggregating agents should be as short as possible
since these enzyme preparations modify the enzymic composition of
the plasma membrane [14]. When the collagenase digest is subjected to rate centrifugation on a linear sucrose gradient in
type A XII or HS zonal rotors (MSE Ltd.), cellular debris, erythrocytes, tubular fragments and glomeruli separate as four distinct zones. Prolonged treatment with collagenase, or homogenization, reduces the length of the tubular fragments recovered but,
with care, extended tubular fragments can be obtained. The properties of commercial collagenase preparations vary from batch to
batch, and often the purest preparations are the least effective
disaggregating agents. Certainly contaminating protease enzymes
play a crucial role, and indeed trypsin which is often present in
collagenase preparations has been used to disrupt kidney tissue
[15, 16].

Isolation of tubular cells

We have found (unpublished data) that short exposure of isolated
tubular fragments to chelating agents such as 0.01 M EDTA results
in the rapid release of cells. By isolating a well defined
nephron region the number of cell types present is limited, and
these regions are a better starting point than the more commonly
used heterogeneous renal cortical cell suspensions.

Disaggregated cells can be separated either by free-flow
electrophoresis or by density-gradient centrifugation. Kreisberg
et al. [17] have exploited the unusually low density of rat-kidney
proximal tubular cells to separate them from other cells by a
combination of isopycnic and rate-zonal centrifugation in Ficoll.
Cells thus isolated from trypsinized rat kidney exhibited high
alkaline phosphatase activity and had prominent brush borders,
and 98% of the cells excluded Trypan Blue. The same research
group [18] also separated rat-kidney cells by electrophoresis, and
the proximal tubular cells were recovered in fractions well
separated from erythrocytes and cells lacking alkaline phosphatase activity. The separation was possible because the alkaline
phosphatase-positive cells had a higher electrophoretic mobility
than the other cells present when they were subjected to free-
flow electrophoresis. A similar electrophoretic technique has
been used (H.-G. Heidrich, this volume) to separate morphologically distinct and viable proximal and distal tubules and renin-
active cells from a single cell suspension prepared from rabbit
renal cortex using citrate [19].

Tissue culture of tubular cells

Isolated renal cells provide a better starting point for culture
than tubular fragments [15] or trypsin-dispersed monkey-kidney
cells [20]. Well established cell lines or short-term cultures
provide material.with which to study the molecular mechanisms
which underly renal function. A persistent problem in the culture
of renal cells is the overgrowth of fibroblasts. The use of a
medium containing D-valine inhibits fibroblast growth and thereby
allows the selective proliferation of epithelial cells [21].
D-valine selected cells retain certain renal functions for several
weeks, and should be of value in the study of renal metabolism.

CONCLUSIONS

1. Biochemical and physiological studies should be carried out
either with intact kidneys or with well defined nephron regions.
2. Glomeruli or tubular fragments rather than whole kidney
should be taken for isolation of viable cells. Such cells provide a convenient starting point for tissue culture.
3. Studies on renal cells can lead to biochemical data which can
be related to the function of individual cell types isolated from
morphologically defined regions of the nephron.

4. Studies using individual cell types will aid in our understanding of normal physiological function and of molecular events in disease.

References

1. Wada, T., Kan, K., Aizawa, K., Kuroda, S., Ino, Y., Inamoto, H., Kitamoto, K., Ogawa, M. & Yamayoshi, W. (1977) *Amer. J. Path. 87,* 323-328.
2. Osathanondh, V. & Potter, E. (1963) *Arch. Path. 76,* 271-276.
3. Guder, W.G. & Schmidt, U. (1974) *Hoppe-Seyler's Z. Physiol. Chem. 355,* 273-278.
4. Schmidt, U. & Guder, W.G. (1976) *Kidney Int. 9,* 233-244.
5. Fong, J.S.C. & Drummond, K.N. (1974) *Meth. Enzymol. 32* [Fleischer, S. & Packer, L., eds.], 653-658.
6. Price, R.G. & Spiro, R.G. (1977) *J. Biol. Chem. 252,* 8597-8602.
7. Grant, M.E., Harwood, R. & Williams, I.F. (1975) *Eur. J. Biochem. 54,* 531-540.
8. Nørgaard, J.O.R. (1976) *Kidney Int. 9,* 278-285.
9. Cohen, M.P. & Vogt, C.A. (1975) *Biochim. Biophys. Acta 393,* 78-87.
10. Freytag, J.W., Ohno, M. & Hudson, B.G. (1976) *Biochem. Biophys. Res. Commun. 72,* 796-802.
11. Burlington, H. & Cronkite, E.P. (1973) *Proc. Soc. Exp. Biol. Med. 142,* 143-149.
12. Camazine, S.M., Ryan, G.B., Unanue, E.R. & Karnovsky, M.J. (1976) *Lab. Invest. 35,* 315-326.
13. Taylor, D.G., Price, R.G. & Robinson, D. (1971) *Biochem. J. 122,* 641-645.
14. Price, R.G., Taylor, D.G. & Robinson, D. (1972) *Biochem. J. 129,* 919-928.
15. Cade-Treyer, D. (1972) *Ann. Inst. Pasteur 122,* 263-282.
16. Younger, J.S. (1954) *Proc. Soc. Exp. Biol. Med. 85,* 202-205.
17. Kreisberg, J.I., Pitts, A.N. & Pretlow, T.G. (1977) *Amer. J. Path. 86,* 591-600.
18. Kreisberg, J.I., Sachs, G., Pretlow, T.G. & McGuire, R.A. (1977) *J. Cell Physiol. 93,* 169-172.
19. Heidrich, H.-G. & Dew, M. (1977) *J. Cell Biol. 74,* 780-788.
20. Cade-Treyer, D. (1975) *Ann. Immunol. 126C,* 201-218.
21. Gilbert, S.F. & Migeon, B.R. (1977) *J. Cell Physiol. 92,* 161-168.

#C-2

ISOLATED CELLS OF GASTROINTESTINAL MUCOSA

C. N. A. TROTMAN
Department of Physiology,
University of Newcastle-upon-Tyne, U.K.
and Department of Biochemistry
   University of Otago, Dunedin, New Zealand

The extensive literature on isolation of gastric and intestinal mucosal cells and separation of sub-populations is surveyed.

Source and type of cell	*Gastric/intestinal mucosa, various species (Table 1): parietal/epithelial (villus, crypt) cells; also colonic tumour, rat.*
Dissociation	*Non-enzymic, usually with chelating agent; or under critical conditions, by whichever protease a particular author finds best (Table 1), possibly with a step to overcome mucous adhesion or DNA aggregation.*
Separation	*Sequential detachment; or rate-sedimentation (usually 1 g) or isopycnically.*
Products	*Morphologically identifiable cells, possibly artefactually spheroid. Viability and functional integrity assessed by biochemical means, notably protein biosynthesis.*
Comments, especially on nomenclature	*Control of bacterial contamination is important. Rat gastric and intestinal cells after isolation continue to synthesize a comprehensive range of different polypeptides for some hours and provide a model system for monitoring acute effects of drugs and other compounds.*
	The term 'isolated' as now employed is synonymous with 'dissociated' as used by some authors, and does not imply homogeneity. [See Preface for nomenclature comments - ED.]

Unlike liver, the vasculature of the gastrointestinal tract does not invite *in vitro* perfusion. Nor is gut clean, sterile, large, compact or free from mucus. Cell isolation techniques are very different from those applicable to hepatic tissue. The field remains one of continuous development, and this brief survey is necessarily selective.

SURVEY OF TECHNIQUES FOR DISSOCIATION OF CELLS FROM TISSUE

Physical and chemical dissociation

Non-enzymic techniques can be used on gastric mucosa from small animals. Bullfrog parietal cells have been isolated selectively in good yield by scraping away successively the surface epithelial cells, then mucous neck cells, then parietal cells [1]. The upper layers were first damaged by incubation with hypertonic (0.5-2.0 M) salt solutions.

The method of Stern [2] is widely quoted and has been modified [3-6]. Rat intestine was incubated at 33° for 11 min with the lumen filled with a buffered solution of salts containing 27 mM sodium citrate. This was replaced with an ice-cold buffered salts solution, or sucrose with buffered salts, and the epithelial cells were detached with manual assistance. Modifications enable cells to be harvested sequentially beginning at the villus tips and extending to the base of the crypts [5-7].

A different approach is to evert the intestine over a rod which is then vibrated at 100 Hz and 2 mm amplitude for 20-30 min [8, 9] or under comparable conditions [10]. The gut is surrounded with saline containing 5 mM EDTA into which villous cells first are shed. In the second phase the sac is inflated with air contained in an inner dialysis tube and briefly re-vibrated to dislodge crypt cells. This technique has been compared favourably [11] with Sjöstrand's [12], in which a machine strips epithelial cells and tissue fragments from rotating everted intestine by the application of mechanical pressure.

Normal and tumourous rat colonic mucosal cells have been isolated by incubation of mucosal scrapings in cold phosphate-buffered salts solution with 200 mM mannitol and antibiotics [13].

Preparation for enzymic dissociation

Vascular perfusion with enzymes is not effective [14]. For all except the smallest animals (mice, rats, small amphibians) the mucosal layer is first dissected from the muscle layer. This confines the preparation to mucosal cells, allows access of enzymes to both sides of the mucosal layer if preferred, and assists oxygenation. Dissection is conveniently done by pulling mucosal and muscle layers in opposite directions with surgical clamps while a scalpel is run between them. The layers can also

be separated by injecting liquid between them [15]. A microscope
slide is convenient for scraping gastric mucosa from small animals. Intestinal mucosa from small animals can be extruded
without opening the intestine, by scraping a slide along the
serosal surface.

For enzymic digestion, the sheet of mucosa may be held flat
at the bottom of an incubation vessel [16], e.g. pinned to wax.
Digestion from one surface permits successive layers to be
harvested sequentially. This facility is lost if the mucosa is
minced, but access of oxygen and of enzyme is improved. The
technique of Lewin *et al*. [17] is to evert a rat stomach [18]
and then to fill this sac with pronase solution which penetrates
from the serosal surface.

Choice enzymic and related conditions

Three principal phases of dissociation are clear [19]. (1) Proteolytic action is directed towards basement membrane and collagen, but the cells remain joined by their junctional complex.
(2) Ca^{2+} depletion causes separation of desmosomes [20] and
possible weakening of tight junctions. (3) Mechanical shearing
forces complete the detachment of remaining adherent regions.

In practice the conditions applicable to a particular species
and tissue are empirically developed but often highly critical.
In order to avoid an anecdotal approach, examples of typical
techniques and selected variables are summarized in Table 1.
Despite the wide range encountered within each variable listed,
the combination of enzyme, concentration, chemical and physical
conditions can be so critical that failure following consistent
success is often traced to a new batch of enzyme or a change in
medium composition. An atypically high yield indicates that
proteolytic activity may have been excessive and accompanied by
cell damage. The choice of enzyme requires experimentation. For
example, in comparisons on *Necturus* gastric mucosa, pronase was
superior to collagenase, hyaluronidase, trypsin and various non-
enzymic methods of isolation [16]; furthermore only Merck pronase was satisfactory. Collagenase was considered best with
guinea-pig [14]. Having achieved success, experimenters sometimes buy large stocks of enzyme from the same batch. Whether
the difference between batches lies in the precise blend of
proteolytic specificities [19, 34], or the presence of activating
or inhibitory metal ions, or a change of enzyme coinciding with
an increase in skill, is much debated. Collagenase requires Ca^{2+}
for activation, therefore with this enzyme the low-Ca^{2+} step is
generally separate.

Incomplete dissociation to single cells may be observed [27]
and has been exploited for isolation of intact gastric glands
with collagenase [23].

Table 1. Examples of gastrointestinal cell isolation procedures with the use of enzymes.

Abbreviations :-
coll = collagenase; pron = pronase; tryp = trypsin; hyal = hyaluronidase; papn = papain.

Notes :-
1. Medium 199 best of several.
2. Gastric glands isolated.
3. Enzymes applied sequentially.
4. Complex procedure; see the ref.
5. The mud-puppy, a tailed amphibian.
6. Pronase on serosal side.
7. Cells fat-loaded *in vivo*.
8. Colon lymphoid cells sought; EDTA/dithiothreitol pre-treatment.

Species	Mucosa	Enzyme	Source & grade	mg/ml	°C	min	O_2	Note	Ref.
Gastric									
rabbit	minced	coll	—	—	37	120	-		21
rabbit	minced	coll	P-L Bio crude	0.8	30	60	-		22
guinea pig	scraped	coll	Sigma Type 1	0.5	37	20 + 20	+	1	14
rabbit	minced	coll	Sigma Type 1	1.0	37	90	+	2	23
dog	minced	pron	Calbiochem B	2.5	25	30	+	3	15
dog	minced	coll	Nutrit. Biochem.	3.75	25	105-135	+		
frog	minced	coll	Sigma	2.0		30	-	3,4	24
frog	epithel	pron	Sigma	2.5		30	-		
Necturus	surface	pron	Merck 70000 U/g	1.75	30	60-120	+	5	16
rat	everted	pron	Merck 70000 U/g	14	37	90	+	6	17
mouse	surface	pron	Merck 70000 U/g	1.5	37	60-77	-		25,26
rat	sac	pron	Kaken 45000 U/g	2-10	36	15-60	-	2	27
rat	sac	tryp	Difco 1 : 250	2-20	36	15-120	-	2	27
Intestinal									
rat	scraped	coll	Sigma Type III	2	37	5-60	-	7	28
human	lamina propria	coll	Worthington CLSPA	0.05	37	18 h	-	4,8	29
rat	intact	pron	Merck 70000 U/g	14	37	40	+	6	30
rat	everted	hyal	Nutrit. Biochem. 300 USP/mg	1.5	ambient	12	+		31
chicken	intact	hyal	—	1.0	37	30	-		32
rabbit	scraped	papn	—	1.0	37	40-45	-		33

   Prepared cells should be washed several times by centrifugation at 100 *g* or higher to remove preparative enzymes. Storage
medium should at least contain buffered salts, glucose and 1-3%
bovine serum albumin (BSA); complex tissue culture media supplemented with BSA are commonly used. Dextran [12, 35] or Ficoll
[11] is occasionally used in place of BSA.

   Mucous adhesion of cells has been tackled with *N*-acetyl-L-
cysteine [2], 1 mM dithiothreitol [29], or repeated washing [22].
If DNA from lysed cells causes aggregation, DNase at 4 µg/ml may
be effective [2].

CELL SEPARATION

The parietal cell has for long been a target for cell purification technology. Apart from the importance and fascination of
this cell, it is larger than most, e.g. 18-25 µm in diameter
after isolation from mouse, compared with 9-15 µm for non-parie-
tal cells [25, 26].

Fractional dissociation: *see above*

Zonal and differential sedimentation at unit gravity

Cells sediment much faster than organelles because, from Stokes'
law, the dependence of sedimentation rate on particle diameter
is second order. The principles are discussed by W.S. Bont &
J.E. de Vries *(this vol.)*.

   For separating gastric mucosal cells we have used 5 or 10 cm
diameter siliconized glass columns containing a gradient of 2-4%
BSA in a minimum essential medium (MEM). Loading and unloading
through a conical top is quicker and causes less disturbance
than through the base. The gradient stabilizes the column
against convection and minor disturbances, as separation is rate-
dependent and not isopycnic. Response to density would be
counter-effective since larger cells are generally less dense.
A cushion of 1.5 M sucrose helps to economize on medium. Cells
survive better in more complete maintenance media than the Krebs-
Ringer type [25]. Pre-warming the medium reduces convection and
prevents formation of bubbles in the tank. Overloading is a
common cause of poor separation; 20 million cells would be a
typical load for a 500 ml gradient in a 10 cm diameter column.

   Gastric glands, being 0.1-0.7 mm in length, were large enough
to be separated from free cells rapidly by pelleting at 1 *g* in
a test tube without a gradient [23].

Accelerated differential sedimentation

Since cells sediment rapidly in dilute aqueous media, they are
centrifuged only for washing or in certain special separation

schemes [14]. Intestinal mucosal cells loaded *in vivo* with corn
oil were separated by flotation from cells of normal density by
centrifugation at 400 *g* [28]. Parietal cells have been partially
purified by centrifugation through 6 → 19% (w/w) Ficoll at 800 *g*
[22], or by repeated differential pelleting at 100 *g* [36].
Centrifugal elutriation (D.L. Knook & E.Ch. Sleyster, *this vol.*)
has also been applied [37].

Isopycnic centrifugation

The behaviour of model cell mixtures in Ficoll gradients in a
zonal rotor has been characterized in detail [38]. Rat gastric
cells have been separated isopycnically in a sucrose gradient
with a zonal rotor [17]. The larger, mainly parietal cells of
mean diameter 16 μm were recovered from the *d* = 1.19-1.20 zone,
and most of the smaller cells of mean diameter 9 μm between 1.24-
1.26. Rat colonic mucosal cells had an apparent density of
1.025 in Ficoll, 1.020 in copper sulphate or sucrose [13]. Ficoll
with Hypaque has been used to separate lymphoid cells, localized
in human colonic mucosa, from the mucosal cells after a dissociation sequence involving EDTA and collagenase [29].

Assessment of fractions

Cell size and number are the criteria usually chosen for the
initial characterization of fractions. Many fractions recovered
from gradients are likely to be too dilute for counting in a
haemocytometer unless first concentrated by pelleting. A Coulter
counter with a particle-size analyzer is ideal. Isolated cells
tend to become spheroid and their dimensions may not correlate
with tissue sections. In dense sucrose media, osmotic contrac-
tion is likely and the indicated isopycnic density is not necessarily related to the value *in vivo*.

Although all cells contain cytochrome oxidase, its abundance
in parietal cells provides a marker for them by assay [17] or
staining [23]. Pepsinogen [17], gastrin, etc. are markers for
their respective cells. The design of separation media is influenced by analytical problems such as interference by sucrose
[39]. Hepes interferes with the Lowry protein assay, as of
course does BSA unless the cells are exhaustively washed. Phase-
contrast and electron microscopy are employed for the recognition
of cells morphologically [17, 25, 36, 40, 41].

VIABILITY

Rapid tests

Rapid screening tests give early but limited information about
the quality of cells. Trypan blue stains dead but not live
cells; an equal volume of 5 mg/ml solution in isotonic medium is
added, then examination in a haemocytometer allows the concentration of cells, the proportion of dead cells, and aggregation

to be estimated. The limitation is that dying cells tend to
exclude the stain until the end, but a departure from the routine
proportion is informative. Eosin [16], Erythrosin [15, 23],
Nigrosin [42] or Lissamine Green V [14] may be used.

Structural integrity

Leakage of cytosol enzymes indicates damaged plasma membrane
(p.m.) and implies that intermediates and cofactors are being
lost. The test entails centrifuging a suspension of cells and
measuring, say, LDH in the supernatant and in the pellet. This
may, however, give a falsely encouraging result, since leaked
enzymes may be inactivated by endogenous or residual added proteases. Membrane permeability can be measured as penetration of
an inert marker, e.g. labelled inulin or sorbitol.

Functional performance

Respiration is easily measured [13, 17, 23, 24, 33, 35, 36], but
indicates only mitochondrial activity, which would not survive
severe damage but would not necessarily disclose limited damage
such as leaky p.m. [see also H.A. Krebs, this vol.]. Damaged
mitochondrial respiratory control could increase respiration.
The value of this parameter again lies chiefly in the detection
of deviant rates. Parietal cells have high oxygen consumption
[15, 40].

Other criteria include active transport of glucose [32, 43],
lactate formation [23, 32], uptake of potassium ions [44, 45],
retention of organic anions [45], and response to relevant
hormones [36, 37, 41, 43] or secretagogues [23, 37, 41, 44].

In view of the limitations of individual tests, cells need to
be characterized by several tests to compile a profile. More
information is needed about an important area of uncertainty,
the possibility of damage by proteolytic enzymes to cell surface
receptors. In pituitary tissue, changes in hormone secretion
have been observed after treatment with pronase at a concentration only 0.1% of that used to dissociate cells [46]. Trypsin
is reported to damage a p.m. transport system for long-chain
fatty acids in liver cells [47].

STERILITY

The support media in which isolated cells are best maintained,
being near neutral, aerobic, and rich in glucose, amino acids
and salts, unfortunately have all the attributes of the best
bacterial growth media. Since at 37° the faster-reproducing
species can increase 10^6-fold in 6 h, moderate contamination
soon dominates biosynthesis [48] and oxygen consumption.
Bacteria also release damaging enzymes [49] and toxins [50].

Avoidance of this problem requires either the total exclusion of all bacteria initially, or their removal from the cell suspension, or inhibition of bacterial activity by antibiotics. Apparatus should be autoclaved, and solutions should be sterilized immediately after preparation, by aseptic filtration through 0.22 μm-pore membranes, or by autoclaving if the constituents are heat-stable.

Whilst most organs can be removed aseptically if necessary, gastrointestinal mucosa scarcely qualifies. Sterilization of the gut *in vivo* is not impossible but demands dedication. Mucosa cannot be freed of bacteria by washing since large numbers inhabit the adherent mucus. Similarly the isolated cells, despite a 10^2-fold higher sedimentation rate than bacteria, cannot be freed of contamination by centrifugation because some types of bacteria attach firmly to the cell surface.

After failure with various antibiotics, we have had no trouble since adopting gentamicin. Among its attributes are broad specificity, a safe margin between working and cytotoxic concentrations, and stability to autoclaving and to a wide range of pH [51]. Cytotoxic concentrations usually exceed 3000 μg/ml [51]. At 1000 μg/ml no acute effect was detected on total protein biosynthesis in our isolated rat gastric cells. Concentrations of 50-200 μg/ml are used in tissue culture [52, 53]. Initially we used 100-200 μg/ml during isolation and experiments with rat cells [30, 35, 48, 54]. This has since been reduced to 10 μg/ml or sometimes 1 μg/ml in order to lessen the possibility of synergism or antagonism with other drugs under study. These concentrations were chosen after examining the effectiveness of gentamicin against bacterial cultures (Table 2).

PROTEIN BIOSYNTHESIS

Protein biosynthesis requires the coordination of several functional compartments of the cell, thus offering a comprehensive assessment of biochemical activity. The technique is basically simple, provided that certain precautions are observed. Cells are incubated with radiolabelled amino acid and then treated with trichloroacetic acid (TCA) to precipitate protein, whose radioactivity is measured. Sterility is crucial since the biosynthetic activity of bacteria is very high. Leucine is the labelled amino acid commonly chosen, being well represented in most proteins. Commercial suppliers offer complex minimum essential media (MEM) containing all essential amino acids except that leucine is specially omitted in order to preserve the specific activity of the isotope. Alternatively an amino acid excluded from the medium as being non-essential, such as proline, can be labelled, but with less incorporation. Higher efficiency of counting and lower dependence on quenching and geometry make ^{14}C worth the extra cost over ^3H.

Table 2. Effectiveness of low concentrations of gentamicin.
Conditions simulated those used to measure protein biosynthesis
in isolated gastric cells. Bacterial cultures were grown aerobically in nutrient broth for 18 h at 37°, then 0.5 ml was transferred to duplicate flasks containing 5 ml of cell support medium
and gentamicin. Labelled leucine was added 30 min later and its
incorporation measured after precipitation with TCA following
3 h incubation at 37° with continuous shaking.

Bacterial culture	$[^{14}C]$leucine incorporation, % of control value, at the stated gentamicin concentration			
	0 (Control)	0.1 µg/ml	1 µg/ml	10 µg/ml
Escherichia coli	100	8.6	0	0
Staphylococcus aureus	100	6.0	0.1	0
Rat gastric mixed culture	100	78	0.4	0.1

Typically, cells are incubated at 37° in flasks containing MEM,
the label and 30 mg BSA/ml. BSA is preferable for short-term
experiments as the free leucine content of serum is high (say
0.2 mM) and inconsistent. Media containing $NaHCO_3$ at a level
exceeding 10 mM as a buffer constituent are intended for use in
a CO_2-enriched atmosphere, but additional buffering with 20 mM
Hepes permits an atmosphere of air. For pH 7.35 at 37°, Hepes
should be adjusted to pH 7.5 at 20°. Conventionally the medium
is buffered at pH 7.3-7.4; the activity of rat gastric cells
peaked at a lower pH (Fig. 1), but this is not necessarily indicative of their physiological environment. Choline (2 mM) or
lysophosphatidylcholine (1.2 mM) when added to Krebs-bicarbonate
incubation medium increased protein biosynthesis by about 30%
in fat-loaded intestinal cells [55]; but bile salts may be strongly
cytotoxic [32].

Unless the cells are washed free from preparative enzymes they
are liable to be short-lived during incubation, and leucine-
specific activity could become diluted by proteolysis of BSA.
Centrifugation four times from 10 ml of support medium has proved
adequate [54]. Division into aliquots before washing gives more
representative aliquots, as freedom from clumps is more difficult
to ensure after washing.

The rate of $[^{14}C]$leucine incorporation deteriorates after 3-6 h
of incubation. Replacement of the medium and label at 3 h and
6 h has been found to restore biosynthesis, suggesting that proteolytic or metabolic dilution of the $[^{14}C]$leucine pool had contributed to the deterioration (Fig. 2).

Fig. 1. Effect of pH on protein biosynthesis by isolated gastric mucosal cells (rat).

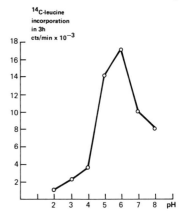

Ideally, dissolved oxygen status should be verified with a mobile oxygen probe. However, knowing the oxygen consumption of the cells, the surface area of the medium and the available shaking rates, the safe maximum load can be estimated [56]. The dependence on shaking rate is steep. As a guide, shaking at 120 oscillations per min with a 2 cm excursion in a flask of liquid surface area 10 cm^2 should permit an uptake of the order of 200 µl or 10 µmol of O_2 per hour from air.

The reaction is stopped and protein is precipitated by addition of cold TCA to 10%. Sodium tungstate [57] or phosphotungstic acid [13] may improve precipitation of small polypeptides. Cold TCA precipitates leu-tRNA; but for many purposes this contribution is neglected. The precipitate is re-suspended and washed four times to minimize leucine adsorption. The BSA present ensures a bulky precipitate and proportionately little loss on washing. Finally the precipitate is dissolved in, say, 0.5 M NaOH at 70° and counted with scintillant.

Necessary controls are a flask to which the label is added after incubation, which corrects for adsorption of label to the precipitate, and a flask with 500 µM cycloheximide, which inhibits eukaryotic but not bacterial activity. The danger of labelled precursor contaminants being incorporated into non-protein macromolecules [58] is checked by dilution of the isotope with pure unlabelled precursor [48, 58]. Incorporation of various other precursors has been studied, e.g. into chylomicrons by isolated rat intestinal cells [55, 59] and into mucoproteins by pig gastric [60] and rabbit intestinal [33] cells.

Specific proteins

In vitro preparations may synthesize predominantly a single protein or type of protein [49, 61]. About 50 newly synthesized labelled proteins have been separated electrophoretically from isolated rat gastric cells incubated for 3 h in medium containing 10 µCi (370 kBq) of [^{14}C]leucine or proline [35]. Incubated cells must be washed substantially free from BSA lest the electrophoresis pattern be distorted. The pellet of cells was extracted by freezing and briefly thawing four times, then homogenizing thoroughly for 2 min at 0° in a small motor-driven glass tissue-grinder.

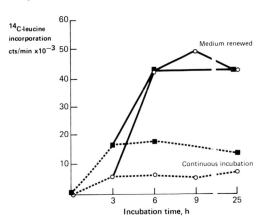

Fig. 2. Effect of medium replacement on protein biosynthesis by isolated gastric mucosal cells (rat) during extended incubation (two experiments, ■ and o). *Continuous line:* cells centrifuged and medium replaced at data points. *Broken line:* original medium retained throughout. Incubation began (t = 0) 2.5 h after removal of the stomach.

Electrophoresis of the 9300 g × 5 min supernatant was conducted in 12.5% polyacrylamide gel containing 0.1% SDS [62] with a Studier [63] apparatus. Treatment with 5% mercaptoethanol and 2% SDS at 100° [62] destroys quaternary (and tertiary) structure of proteins so that subunits are detected, and electrophoretic mobility is related inversely to the logarithm of molecular weight (deviations occur at low molecular weights, or with glyco-proteins). Proteins stained with Coomassie Blue R250 were recorded by photography and densitometry of the wet gel (Fig. 3).

Radioactivity was detected by fluorography of the gel [64]. This involves impregnation with the scintillant 2,5-diphenyl-oxazole (PPO), drying, then exposing blue-sensitive film to the gel to reveal an image of the protein bands proportional to their radioactivity. The film can be scanned densitometrically (Fig.3). The advantage of fluorography over autoradiography is its higher sensitivity, particularly if photographic exposure is done at -70° [citation in 64]. Unlike autoradiography, the response to radioactivity can be made quantitative by pre-exposing the film [64]. The disadvantage is that stained bands are largely obscured by the PPO. Matching a large number of fluorographically-detected bands to the photograph of the stained gel is not always unambiguous since slightly irregular shrinkage can occur during impregnation and drying. Some direct autoradiographs of dried stained gels were therefore prepared on radiographic film for arbitrating any slight disparities between fluorographic and stained bands.

Quantitative densitometry

Quantitative monitoring of a particular protein band really calls for integration of its area on the densitometric record, which is difficult since the bands overlap extensively. Computer programs are available based on various mathematical approaches for the resolution of overlapping Gaussian peaks, listing their

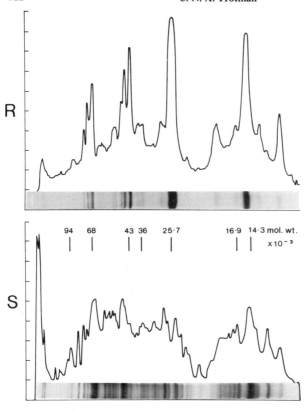

R

94 68   43 36   25·7   16·9 14·3 mol. wt.
| | | | | | | x10⁻³

S

Fig. 3. Bio-
synthesis of
proteins labelled
with [¹⁴C]proline
by isolated
gastric mucosal
cells (rat). The
photographs show
the proteins de-
tected by stain-
ing (S) on SDS-
polyacrylamide
electrophoresis
gel, and radio-
activity (R) de-
tected on the
same gel by
fluorography.
Corresponding
densitometric
recordings (arbi-
trary scales) are
shown above.
Molecular weights
of marker
proteins are in-
dicated. Migra-
tion was from
left (cathode)
to right.

individual areas and statistical characteristics [65, 66]. Experimentally induced changes
in a band can thus be quantified more accurately than from the
height of the trace. An interactive program also enables a pair
of traces to be resolved together and the ratios of all corresponding peaks in the two traces to be listed [65].

     Important limitations govern the validity of comparisons
between different stained gels, or fluorographic films, but most
are in theory avoided by normalizing all individual band areas
to the total area under the trace of the complete gel track.
Otherwise, differences may originate in the gels being treated
unequally in staining, impregnation with scintillant, film exposure time and temperature, film sensitivity and also development. Normalization also minimizes experimental differences in
cell concentration, labelling efficiency, extraction and gel
loading. Other variables cannot easily be corrected for. A
protein stains quantitatively within limits [67, 68]; but different proteins do not necessarily stain similarly, nor necessarily
label similarly owing to differences in content of leucine or
other label.

Relative specific radioactivity

Relative specific radioactivity of a band is calculated as the normalized area of the fluorographic peak divided by that of the stained peak. A suitable program [65] compares the resolved areas automatically, provided that the gels are synchronized. Experimental changes in the relative specific radioactivity of a single peak are clearly of more quantitative significance than comparisons between different peaks.

APPLICATIONS

In view of the present emphasis on methodology, illustrative examples will merely be touched on. One application of the studies of protein biosynthesis is the feedback information obtained about the viability and functional competence of isolated cells. Our experiments with rat gastric cells show the following:-
1. Thorough washing to remove pronase is essential [54]. 2. The cells may continue *in vitro* for at least 8 h to synthesize a wide range of soluble protein molecules (Fig. 2). 3. Generally these new proteins correspond to existing proteins (Fig. 3). 4. Labelling need not correlate quantitatively with stain density, and newly synthesized proteins therefore differ in relative specific radioactivity, suggesting differential synthesis or turnover [35]. 5. Profiles of staining and radiolabelling are reproducible in different batches of similar cells. These observations together suggest that isolated cells retain a controlled and regulated system for the biosynthesis of polypeptides.

In cell separation experiments with rat gastric mucosa we have established marked differences in biosynthesis profiles; for example the smaller cells (predominantly <800 μm^3, 11.5 μm) actively synthesize a 25000 mol. wt. molecule, the larger cells one of 43000 mol. wt.

Gastrointestinal mucosa *in vivo* may suffer non-physiological treatment, e.g. its use as a route for the delivery of drugs. A tablet or capsule containing the calculated dose for the whole body may lodge in the mucus layer and deliver a vastly higher concentration to cells locally. Antibiotics are obvious candidates for drugs that, at excessive local concentrations, may damage eukaryotic biosynthesis. We have found clindamycin at 4 mg/ml and the related compound lincomycin at 20 mg/ml to halve biosynthesis over 3 h in rat gastric cells; pharmaceutical capsules contain 150 mg and 500 mg respectively. Both drugs disturbed synthesis of 47000, 43000 and 25000 nominal mol. wt. molecules. Clioquinol and the related 8-hydroxyquinoline inhibited biosynthesis by 50% at 3.5 mM. Tetracycline, to which eukaryotic cells are regarded as impermeable, had no effect at up to 2 mg/ml, although effects on intestinal protein biosynthesis after treatment *in vivo* have been reported [69]. Inhibitory effects of salicylate have been observed [70, 71].

In principle, isolated and separated target cells are ideal for investigation of the effects of secretagogues and modulators [37, 44, 72]; isolated gastric glands have also been rewarding here [41, 73].   Numerous studies have centred on parietal cell function [1, 14, 15, 17, 22-26, 37, 41, 72, 73].

Fatty acid uptake [32] and iron absorption [10] have been studied in isolated intestinal cells. Intestinal cells separated after fat-loading have proved valuable in studies of chylomicron synthesis and transport [28, 55, 59] and of the role of biliary phosphatidylcholine [55]. Intestinal cells infected with the protozoan *Eimeria necatrix* have been studied at different stages of infection after isolation on the basis of consequent size and other differences [74].

Acknowledgements

The author's work was supported by grants from the Royal Society, the Medical Research Council and the Smith, Kline & French Foundation. The assistance of Miss Christina Spink and Miss Hilary Duckworth was indispensable. Mrs. Eleanor M.S. Harte provided a most professional and efficient scientific information service.

References
1. Forte, J.G., Ray, T.K. & Poulter, J.L. (1972) *J. Appl. Physiol. 32,* 714-717.
2. Stern, B.K. (1966) *Gastroenterol. 51,* 855-864.
3. Douglas, A.P., Kerley, R. & Isselbacher, K.J. (1972) *Biochem. J. 128,* 1329-1338.
4. Eastham, E.J., Bell, J.I. & Douglas, A.P. (1977) *Biochem. J. 164,* 289-294.
5. Weiser, M.M. (1973) *J. Biol. Chem. 248,* 2536-2541.
6. Charney, A.N., Gots, R.E. & Gianella, R.A. (1974) *Biochim. Biophys. Acta 367,* 265-270.
7. Raul, F., Simon, P., Kedinger, M. & Haffen, K. (1977) *Cell Tiss. Res. 176,* 167-178.
8. Harrison, D.D. & Webster, H.L. (1969)   *Exp. Cell Res. 55,* 257-260.
9. Webster, H.L. & Harrison, D.D. (1969)   *Exp. Cell Res. 56,* 245-253.
10. Halliday, J.W. & Powell, L.W. (1973) *Clin. Chim. Acta 43,* 267-276.
11. Iemhoff, W.G.J., Van den Berg, J.W.O., de Pijper, A.M. & Hülsmann, W.C. (1970) *Biochim. Biophys. Acta 215,* 229-241.
12. Sjöstrand, F.S. (1968)   *J. Ultrastruct. Res. 22,* 424-442.
13. Perret, V., Lev, R. & Pigman, W. (1977) *Gut 18,* 382-385.
14. Jewell, D.P., Katiyar, V.N., Rees, C., Taylor, K.B. & Wright, J.P. (1975) *Gut 16,* 603-612.
15. Croft, D.M. & Ingelfinger, F.J. (1969) *Clin. Sci.37,* 491-501.

16. Blum, A.L., Shah, G.T., Wiebelhaus, V.D., Brennan, F.T., Helander, H.F., Ceballos, R. & Sachs, G. (1971) *Gastroenterol. 61*, 189-200.
17. Lewin, M., Cheret, A.M., Soumarmon, A. & Girodet, J. (1974) *Biol. Gastroenterol. (Paris) 7*, 139-144.
18. Dikstein, S. & Sulman, F.G. (1965) *Biochem. Pharmacol. 14*, 335-357.
19. Amsterdam, A. & Jamieson, J.D. (1974) *J. Cell Biol. 63*, 1037-1056.
20. Sedar, A.W. & Forte, J.G. (1964) *J. Cell Biol. 22*, 173-188.
21. Walder, A.I. & Lunseth, J.B. (1963) *Proc. Soc. Exp. Biol. Med. 112*, 494-496.
22. Glick, D.M. (1974) *Biochem. Pharmacol. 23*, 3283-3288.
23. Berglindh, T. & Öbrink, K.J. (1976) *Acta Physiol. Scand. 96*, 150-159.
24. Kasbekar, D.K. & Blumenthal, G.H. (1977) *Gastroenterol. 73*, 881-886.
25. Romrell, L.J., Coppe, M.R., Munro, D.R. & Ito, S. (1975) *J. Cell Biol. 65*, 428-438.
26. Munro, D.R., Romrell, L.J., Coppe, M.R. & Ito, S. (1975) *Exp. Cell Res. 96*, 69-76.
27. Kurokawa, Y., Saito, S., Kanamaru, R., Sato, T. & Sato, H. (1975) *Tohoku J. Exp. Med. 116*, 241-252.
28. Yousef, I.M. & Kuksis, A. (1972) *Lipids 7*, 380-388.
29. Bull, D.M. & Bookman, M.A. (1977) *J. Clin. Invest. 59*,966-974.
30. Trotman, C.N.A. (1978) *Biochem. Soc. Trans. 6*, 626-628.
31. Perris, A.D. (1966) *Canad. J. Biochem. 44*, 687-693.
32. Haag, G., Bierbach, H. & Holldorf, A.W. (1976) in *Lipid Absorption: Biochemical and Clinical Aspects* (Rommel, K. & Bohmer, R., eds.), MTP Press, Lancaster, pp. 335-340.
33. Padron, J., Gallagher, J.T. & Kent, P.W. (1973) *Brit. J. Exp. Path. 54*, 347-351.
34. Amsterdam, A. & Jamieson, J.D. (1972) *Proc. Natl. Acad. Sci. U.S.A. 69*, 3028-3032.
35. Trotman, C.N.A., Sanderson, C. & Spink, C. (1978) *Cell Biol. Int. Rep. 2*, 177-184.
36. Soumarmon, A., Cheret, A.M. & Lewin, M.J.M. (1977) *Gastroenterol. 73*, 900-903.
37. Soll, A.H. (1977) *Gastroenterol. 73*, 899.
38. Boone, C.W., Harell, G.S. & Bond, H.E. (1968) *J. Cell Biol. 36*, 369-378.
39. Hartman, G.C., Black, N., Sinclair, R.&Hinton,R.H. (1973) in *Subcellular Studies* [Vol. 4, *this series*] (Reid, E., ed.), Longman, London, pp. 93-102.
40. Ito, S., Munro, D.R. & Schofield, G.C. (1977) *Gastroenterol. 73*, 887-898.
41. Berglindh, T., Helander, H.F. & Öbrink, K.J. (1976) *Acta Physiol. Scand. 97*, 401-414.
42. Kaltenbach, J.P., Kaltenbach, M.H. & Lyons, W.B. (1958) *Exp. Cell Res. 15*, 112-117.
43. Thieden, H.I.D., Quistorff, B., Selmer, J. & Grunnet, N. (1976)

in *Use of Isolated Liver Cells and Kidney Tubules in Metabolic Studies* (Tager, J.M., Söling, H.D. & Williamson, J.R., eds.), North-Holland, Amsterdam, pp. 233-244.

44. Batzri, S. (1977) *Gastroenterol. 73*, 913.
45. Quistorff, B., Bondesen, S. & Grunnet, N. (1973) *Biochim. Biophys. Acta 320*, 503-516.
46. Schofield, J.G. & Orci, L. (1975) *J. Cell Biol. 65*, 223-227.
47. Mahadevan, S. & Sauer, F. (1974) *Arch. Biochem. Biophys. 164*, 185-193.
48. Sanderson, C. & Trotman, C.N.A. (1977) in *Membranous Elements and Movement of Molecules* [Vol. 6, *this series*] (Reid, E., ed.), Horwood, Chichester, pp. 389-391.
49. Allen, A. & Starkey, B.J. (1974) *Biochim. Biophys. Acta 338*, 364-368.
50. Beeken, W.L., Roessner, K.D. & Krawitt, E.L. (1974) *Gastroenterol. 66*, 998-1004.
51. Schafer, T.W., Pascale, A., Shimonaski, G. & Came, P.E. (1972) *Appl. Microbiol. 23*, 565-570.
52. Casemore, D.P. (1967) *J. Clin. Path. 20*, 298-299.
53. Clancy, R. (1976) *Gastroenterol. 70*, 177-180.
54. Trotman, C.N.A. (1977) *Cell Biol. Int. Rep. 1*, 541-547.
55. O'Doherty, P.J.A., Yousef, I.M., Kakis, G. & Kuksis, A. (1975) *Arch. Biochem. Biophys. 169*, 252-261.
56. Umbreit, W.W., Burris, R.H. & Stauffer, J.F. (1959) *Manometric Techniques, 1st edn.*, Burgess, Minneapolis.
57. Griffin, A.C., Ward, V., Canning, L.C. & Holland, B.H. (1964) *Biochem. Biophys. Res. Commun. 15*, 519-524.
58. Oldham, K.G. (1971) *Anal. Biochem. 44*, 143-153.
59. O'Doherty, P.J.A., Yousef, I.M. & Kuksis, A. (1973) *Arch. Biochem. Biophys. 156*, 586-594.
60. Snary, D. & Allen, A. (1972) *Biochem. J. 127*, 577-587.
61. Sutton, D.R. & Donaldson, M. (1975) *Gastroenterol.69,* 166-174.
62. Laemmli, U.K. (1970) *Nature (Lond.) 277*, 680-685.
63. Studier, F.W. (1973) *J. Mol. Biol. 79*, 237-248.
64. Laskey, R.A. & Mills, A.D. (1975) *Eur. J. Biochem. 56,* 335-341.
65. Trotman, C.N.A. & Greenwell, J.R. (1979) *Biochem. J. 178*, 159-164.
66. Booth, A.G. & Kenny, A.J. (1976) *Biochem. J. 159*, 395-407.
67. Condeelis, J.S. (1977) *Anal. Biochem. 77*, 195-207.
68. Weber, K., Pringle, J.R. & Osborn, M. (1972) *Methods in Enzymol. 26C*, 3-27.
69. Greenberger, N.J. (1967) *Nature (Lond.) 214*, 702-703.
70. Rainsford, K.D. & Smith, M.J.H. (1969) *Biochem. J. 111*, 37P.
71. Kent, P.W. & Allen, A. (1968) *Biochem. J. 106*, 645-658.
72. Lewin, M., Cheret, A.M., Soumarmon, A., Girodet, J., Ghesquier, D., Grelac, F. & Bonfils, S. (1976) in *Stimulus Secretion Coupling in the Gastrointestinal Tract* (Case, R.M. & Goebell, H., eds.), MTP Press, Lancaster, pp. 371-375.
73. Berglindh, T. (1977) *Acta Physiol. Scand. 99*, 75-84.
74. Fernando, M.A. & Pasternak, J. (1977) *Parasitology 74*, 19-26.

#C-3
PREPARATION OF DUCT CELLS FROM THE PANCREAS

I. SCHULZ, K. HEIL, S. MILUTINOVIĆ, W. HAASE,
D. TERREROS and G. RUMRICH
Max-Planck-Institut für Biophysik,
Kennedyallee 70, Frankfurt/Main, W. Germany

Pancreatic acinar cells are easily obtainable [1]. The following method provides duct cells in satisfactory amount and purity.

Source — *Pancreas, young rats, 0.1 g scale.*

Dissociation — *A 4-step procedure [1] entailing two enzymic steps with intervening EDTA and CaCl$_2$ treatments.*

Separation — *Sedimentation at 50 g through a 4% albumin gradient, pelleting of the heavier acinar cells, collection of the lighter cells remaining in the supernatant, and purification of duct cells, e.g. with an elutriator rotor.*

Product — *Recovery 0.9% of pancreatic cell protein, mainly (80-90%) duct cells; 10-fold enrichment of Na$^+$K$^+$-stimulated ATPase as compared with homogenate from the total organ.*

Comment and alternative — *In the purification step for duct-cells, elutriation works better than centrifugation in a Ficoll density gradient. With tissue from 50 rats the yield of duct cells is $\sim$250 mg, allowing isolation of plasma membranes with a zonal rotor [2].*

Much progress in the understanding of enzyme synthesis and secretion from the exocrine pancreas as well as of insulin release from the endocrine pancreas has been made through the availability of functioning isolated acinar cells [1] and intact islets (G. Raydt, *this vol.*). The function of isolated duct cells, however, has not received comparable attention, no doubt due to the difficulty of purifying these cells in satisfactory amount. Isolation of duct cells has been carried out by three different procedures. —
1. Microdissection of ducts from the total organ *in vitro* [3].
2. Enzymatic digestion of the total organ with collagenase and hyaluronidase [1] and subsequent purification of dissociated cells

on a Ficoll density gradient or by counter-flow centrifugation.
3. Selective destruction of acinar cells by a copper-deficient
diet supplemented with D-penicillamine leaving duct cells intact
[4].

All three methods furnish preparations of cells virtually
free of zymogen granules. However, since duct cells in the pancreas comprise only 2-4% of the total tissue, the main limitation of separation methods is the low yield of duct cells. We
now describe two approaches which were undertaken in our laboratory to isolate duct cells from the pancreas. By microdissection
medium and large size ducts could be obtained, whereas the smallest ducts and centro-acinar cells, both considered to be especially active in bicarbonate and fluid secretion, could not be dissected. Furthermore, the yield of duct cells gained by microdissection was small (less than 0.1% of the total pancreas, wet
wt.). Enzymatic digestion of the total tissue with collagenase
and hyaluronidase and subsequent separation of duct cells in a
Beckman elutriator rotor leads to a much better yield (0.2-0.5%
of duct-cell protein from total tissue protein). With the latter
method, recovery of duct cells smaller than 10 µm in diameter is
possible.

MICRODISSECTION OF DUCT FRAGMENTS FROM RAT PANCREAS

The pancreas was removed from the rat after exsanguination of the
animal and kept in ice-cold Krebs-Ringer bicarbonate (KRB) buffer,
oxygenated with 95% O_2, 5% CO_2. Pancreatic ducts of different
diameters (50 to 1000 µm) were dissected from the pancreas on
ice using a stereomicroscope Wild M 5 at 25 x and 50 x magnification, light being transmitted from below. Fine-tip Dumont forceps and scissors for eye-surgery were used for dissection. Ducts
were prepared starting from the main duct and following its
branches up to ∿50 µm (Plates 1 and 2). Smaller ductlets especially those emerging from the acini could not be obtained with
this method, nor centro-acinar cells. From 8 rats ∿6 mg of
ducts (wet weight) were obtained by one person after 1 h of dissection, which is ∿0.08% of total tissue wt. Since ducts are
surrounded by several layers of connective tissue (Plate 1), experiments in which the intracellular space has to be known pre-
cisely turned out to be difficult. Extracellular water space
made up 40-80% of the total tissue wet wt. as measured with ^{3}H-
or ^{14}C-inulin, ^{3}H- or ^{14}C-polyethylene glycol, ^{3}H- or ^{14}C-dextran
or ^{3}H-mannitol after centrifugation of ducts at 10,000 g for
2 min in an Eppendorf centrifuge to obtain a tightly packed
pellet.

The method described by Fölsch and Creutzfeldt [4] whereby
acinar cells are destroyed selectively in rats given a copper-
deficient diet plus D-penicillamine (Plate 3) has been tried by

Plate 1. Electron micrograph (likewise Plates 2-5). An isolated pancreatic duct showing epithelial duct cells (D) in the upper part, and lumen (L). A thick layer of connective tissue, containing blood vessels and small ducts (SD), surrounds the epithelial cell layer. x 1100.

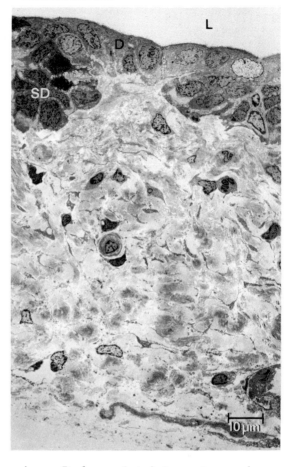

us to obtain pancreases free of acini as starting material for duct dissection. However, since we did not find it easier to dissect ducts from such tissues, we have developed a method which employs dispersion of pancreatic cells by enzymatic digestion and subsequent separation of duct cells by either counter-flow technique using a Beckman elutriator rotor or by centrifugation of cells on a Ficoll density gradient.

PREPARATION OF DISPERSED PANCREAS CELLS

Pancreases from 12-16 rats were artificially perfused *in situ* from the portal vein with a modified KRB buffer (0.125 mM $CaCl_2$) in order to remove blood cells. The KRB buffer contained the following additions: Na^+ fumarate (5 mM), Na^+ pyruvate (5 mM), Na-α-ketoglutarate (5 mM), glucose (15 mM) and, each 1% of the final volume, 'Basal medium (Eagle) amino acids' (100 times concentrated) and 'Basal medium (Eagle)' vitamin solution (100 times concentrated) from Grand Island Biological Co., Grand Island, New York. Pancreatic cells were isolated according to the method described by Amsterdam & Jamieson [1] with some modifications. Briefly it involves:
Step 1): Collagenase (Sigma, Type I) 1.50 mg/ml; hyaluronidase

Plate 2. Epithelial cells of a rat pancreatic duct. The cell surface at the lumen (L) is characterized by microvilli. Laterally the cells are linked by small microvilli-like cytoplasmic extrusions. The nucleus (N) is elongated and shows indentations. The cytoplasm contains mitochondria (M) and a poorly developed e.r. x 8000.

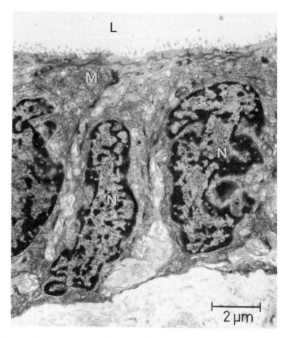

(Boehringer) 3.0 mg/ml, soybean trypsin inhibitor (T.I., Boehringer) 0.2 mg/ml in the modified KRB buffer; shake 15 min at 37°.
Step 2): EDTA 2 mM, T.I. 0.2 mg/ml in KRB buffer without $CaCl_2$; shake 2 × 5 min.
Step 3): $CaCl_2$ 0.125 mM, T.I. 0.2 mg/ml in KRB buffer; shake 2 × 1 min.
Step 4): Collagenase 2.5 mg/ml, hyaluronidase 4 mg/ml, T.I. 0.2 mg/ml in KRB buffer with 0.125 mM $CaCl_2$, shake 45 min at 37°.
The digestion medium was discarded and the tissue washed 3 times with 80 ml of the modified KRB buffer. The pancreases were dispersed by passing them 2 times through a plastic pipette tip ('Pipetman', Gilson) of 5 mm bore, then about 5 times through a series of pipette tips with decreasing bore (5 mm, 4 mm, 3 mm, 2 mm) and about 20 times with a pipette tip of 0.7 mm.

Heavier acinar cells were separated by sedimentation at 10 g for 3 min through an albumin cushion: 13 ml of cell suspension was layered over 9 ml of 4% albumin (w/w) in KRB buffer. Lighter cells remaining in the supernatant, including acinar and duct cells, were further purified by counter-flow centrifugation.

SEPARATION OF DUCT CELLS IN THE BECKMAN ELUTRIATOR ROTOR

The principle of use of the Beckman elutriator rotor to separate cells is shown in Scheme 1. Cell separation takes place in the

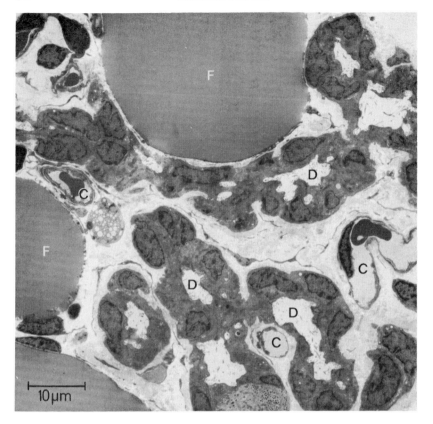

Plate 3. A rat pancreas after 11 weeks on a copper-deficient
diet containing D-penicillamine. The exocrine pancreas consists
predominantly of ducts (D), blood capillaries (C), and huge fat
vacuoles (F) embedded in connective tissue. No acinar cells
could be detected. x 1700.

separation chamber by counter-action of two forces; viz. centrifugal force which can be adjusted by the speed of rotor revolu-
tion, and a counterflow adjusted by an LKB Multiperpex pump.
The chamber of the running rotor can be observed through a win-
dow in the lid of the centrifuge by means of an in-built stro-
boscope.

Cells suspended in KRB buffer, containing 0.1 mM $CaCl_2$ and
1.6% albumin, were loaded at a flow rate of 61 ml/min into the
rotor running at a speed of 2000 rev/min at 4°. At this rate
groups of acini, bigger isolated acinar cells, and only a few
per-cent of non-granulated cells remained in the chamber. The
cells could be removed from the chamber either by increasing the

Scheme 1.
Principle of
the Beckman
elutriator
rotor. For
explanation
see text (cf.
D.L. Knook &
Elizabeth Ch.
Sleyster,
this vol.).

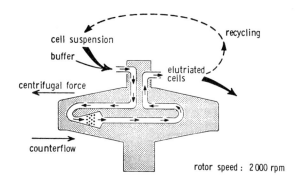

	flow rate [ml / min]	separated cells in the chamber
1. load	61	groups of acini ; duct cells
2. load	29	big isolated acinar cells ; duct cells
3. load	21.5	small acinar cells ; duct cells
4. load	18.5	duct cells ; acinar cells
5. load	15	duct cells ; acinar cells
6. load	12.2	duct cells ;· no acinar cells
7. load	5.4	duct cells ; no acinar cells

Between each load : a) wash until effluent buffer is free of cells
b) harvest remaining cells from the chamber

flow-rate or by direct removal from the chamber after having
stopped both the pump and the rotor. We preferred the latter
since the cells were obtained in a concentrated suspension. The
cells which had passed the chamber before were loaded again on
the rotor, but with a smaller flow rate (29 ml/min) at which
mainly big acinar cells were found in the chamber. Subsequent
loading with smaller pump rates then yielded smaller acinar
cells and more cells without granules. At the 6th load (12.1
ml/min) and the 7th load (5.4 ml/min) only cells without
granules, which showed typical characteristics of duct cells,
remained in the chamber (Plate 4).

PURIFICATION ON A FICOLL DENSITY GRADIENT

Alternatively to elutriation duct cells were purified by centrifugation through a Ficoll density gradient. The supernatant obtained from the 4% albumin gradient *(see above)* was collected
and diluted in 10 volumes of KRB buffer plus 0.1 mg/ml T.I. The

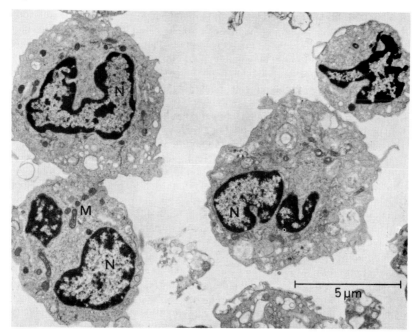

Plate 4. Isolated rat duct cells obtained from the elutriator
rotor. Cells have become rounded off during isolation and show
some degree of vacuolization in the cytoplasm. The indentations
of the nuclei (N) are still present; mitochondria (M) are well
preserved. x 6800.

cell suspension was centrifuged in 20 ml centrifuge tubes in a
Christ Omega centrifuge at 150 g for 1.5 min. The resulting
supernatant was centrifuged at 340 g for 2 min and the supernatant obtained by this step at 600 g for 2 min. All three
pellets were taken up in KRB buffer + 1 mg/ml T.I. and combined
to make up 2 ml. The cells were layered over a Ficoll step gradient consisting of 2 ml of each 1%, 2%, 3%, 4%, 5%, 6%, 7% and
8% Ficoll in KRB buffer with increasing Ficoll concentrations
from the top to the bottom of the centrifuge tube. The gradient
was centrifuged at 186 g for 10 min at 4° in a Beckman swing-out
rotor SW 27. Duct cells were collected between 3% and 5% Ficoll
(Plate 5), whereas acinar cells were found in the pellet.

MORPHOLOGY OF DUCT CELLS

For examination of cell morphology, tissue and isolated cells
were fixed with 2.5% glutaraldehyde buffered with 0.1 M sodium
cacodylate (pH 7.4) and post-fixed (1% OsO_4, 0.1 M cacodylate
buffer, pH 7.4). Dehydrated material was embedded in a plastic

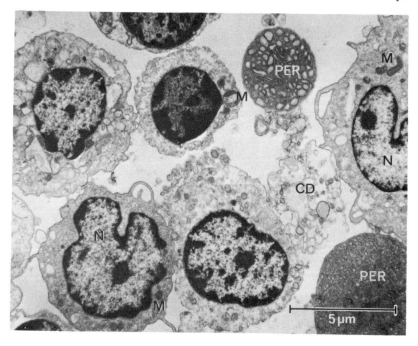

Plate 5.   Isolated duct cells obtained from the Ficoll gradient.
The cells are of globular shape. The cytoplasm contains partly
vacuolized e.r., indentated nuclei (N) and distinct mitochondria (M). Portions of e.r. derived from acinar cells (PER) and
cell debris (CD) are present. x 6800.

resin (Spurr's medium) and sectioned with an LKW ultramictrotome.
Sections were double-stained with uranyl acetate and lead citrate
and examined in a Philips 300 electron microscope.

Ducts obtained by microdissection show cells 6-8 µm in diameter, which is about 2-3 times smaller than that of acinar cells.
It can be seen that ducts are surrounded by connective tissue
which is 10 times as thick as the single duct cell layer (Plates
1 and 2).

Duct cells recovered in fractions 6 and 7 from the elutriator
(Plate 4) and from the Ficoll gradient (Plate 5) show morphological characteristics of duct cells. The most prominent feature
is their big nucleus which is round or contains indentations.
Zymogen granules, which are characteristic for acinar cells, are
absent. The endoplasmic reticulum (e.r.) is poorly developed
and only a few mitochondria are present.

DISCUSSION

Our results indicate that the method successfully applied for the dissociation of pancreatic acinar cells [1] also proves useful for isolated duct cells. Duct cells separated in a Ficoll gradient showed the same morphology as those obtained from the elutriator rotor (Plates 4 & 5), but the yield using a swing-out bucket rotor is limited, and only collecting from several experiments could yield sufficient material for biochemical studies. With the elutriator rotor it is possible to obtain cells continuously and therefore, in principle, unlimited amounts of material could be separated. The limiting factor is the starting material and the time; the cells remain in good condition. In our procedure we separated a total of 10 g of isolated cells within 2 h on the elutriator rotor to obtain 7 fractions; two of them containing duct cells made up 2-4 mg of protein which represents a total protein recovery of 0.2-0.5%. The purity of cells without granules was 90-95% as checked by light and electron microscopy. The rest of the cells were erythrocytes. Using the Ficoll gradient, only 0.05-0.15% recovery of duct-cell protein from total-cell protein could be obtained. We also found that separation was much sharper using the elutriator rotor than with the Ficoll gradient. Especially small particles, e.g. broken cells and cell bodies as always present in fractions from the Ficoll gradient, could be washed away in the counter-flow from intact cells nearly completely. We had not made any attempts to separate remaining red blood cells from duct cells, but this should be feasible by loading the elutriator rotor with the fraction containing duct cells and modifying the flow rate in smaller steps than those used for separation of 20 μm-diameter acinar cells from 10 μm-diameter duct cells.

The method of enzymatic digestion with subsequent purification of duct cells proved to be superior to microdissection in respect of yield of duct cells, especially of small ones considered to be most active in $NaHCO_3$ and fluid secretion.

The availability of purified duct cells should facilitate studies on pancreatic transport processes that take place in the duct system, e.g. intracellular pH-measurements in response to secretin. This is under current investigation in our laboratory.

References

1. Amsterdam, A. & Jamieson, J.D. (1972) *Proc. Nat. Acad. Sci. U.S.A. 69,* 3028-3032.
2. Milutinović, S., Sachs, G., Haase, W. & Schulz, I. (1977) *J. Membrane Biol. 36,* 253-279.
3. Wizemann, V., Christian, A.-L., Wiechmann, J. & Schulz, I. (1974) *Pflügers Arch. 347,* 39-47.
4. Fölsch, U.R. & Creutzfeldt, W. (1977) *Gastroenterology 73,* 1053-1059.

#C–4
CENTRIFUGAL ISOLATION OF PANCREATIC ISLETS IN METRIZAMIDE

GERHARD RAYDT

Institut für Physiologische Chemie, Physikalische
Biochemie und Zellbiologie der Universität München,
München 2, W. Germany.*

Source of material	Pancreases from 150-200 g Sprague-Dawley rats having free access to Altromin standard diet.
Dissociation	Digestion with 20 mg collagenase [1], 20 min, 37°, in 10 ml Hank's albumin solution (0.2 g BSA and 1.0 g glucose/l) with magnetic stirring, 3 x 5 min sedimentation at 1 x g to remove the bulk of the digested exocrine pancreas with the supernatant.
Separation	The last sediment is mixed with 30% w/v Metrizamide[†] in Hank's albumin solution to give a refractive index of 1,357, and layered under an isopycnic Metrizamide-Hank's albumin step gradient with refractive indexes as stated; after the run the centrifuge is stopped without braking in order to avoid vortexing of the gradient. Alternatively one can use a linear Metrizamide gradient (1.356 to 1.352), or (see below) other gradient media.
Product	Near-quantitative recovery determined by stereomicroscopic examination of the sediment. For purity and viability, see below.

Islets thus isolated are less contaminated with exocrine
tissue, judged by specific α-amylase activity and by phase-
contrast microscopy, than islets purified 3-fold by hand under
the stereomicroscope. The glucose-mediated stimulation of insulin secretion and ^{3}H-uridine incorporation into total cellular
RNA was preserved, implying that the gradient method gives functionally intact islets in high purity with little effort compared

---

*PRESENT ADDRESS: Augenklinik der Universität, München 2, Mathildenstr.
[†]3,5-Diiodo-4-pyridone-N-acetic acid methyl-glucamine salt,
obtainable from Nyegaard, Oslo

*with isolation by hand. Thus islets can be purified in quantity
for biochemical studies and possibly for transplantation experiments. Consideration is given to induction of transplantable
tumours of one cell type, and to cultivation of islets in vitro,
which could lead ultimately to pure cell lines of endocrine
cells and to artificial transplantable organs.*

The main obstacle to biochemical examination of islet structure
and function is the scarcity of material. After collagenase
digestion of the pancreas according to the established method
of Lacy and Kostianovky [1] the islets have to be collected by
hand under the stereomicroscope. Several authors tried to
circumvent this tedious work by using preparative gradients for
the separation of the islets from the exocrine tissue. Lacy and
Kostianovsky [1] described a sucrose step gradient, and Gerner
et al. [2] used a Ficoll gradient. The glucose-mediated stimulation of insulin secretion by the islets thus prepared was,
however, reduced compared to islets collected by hand. This was
ascribed to some detrimental effect of the gradient media on
the islets [1, 3].

When I started to purify islets, reports had appeared on the
successful iso-osmotic separation of blood and liver cells and
cultured cells in Metrizamide gradients without toxic damage
[4]. So I applied this method to the isolation of islets from
rat pancreas after collagenase digestion. The scheme on p. 142
outlines the finally adopted procedure, as published in detail
[5].

COMMENTS ON THE PROCEDURE AND THE PRODUCT

The islets ascend to the position of their buoyant density, and
the exocrine tissue sediments to the bottom of the tube. Some
fat tissue bound to excretory and connective tissue ascends to
the Metrizamide-Hank's boundary. The islets are carefully aspi-
rated with a Pasteur pipette. They are still contaminated by
some blood vessels. These are removed by quickly collecting the
islets under the stereomicroscope or by velocity sedimentation
in Metrizamide-Hank's albumin solution with a refractive index
of 1.3525 at $1 \times g$ for 2-5 min.

The purity of the islets, the ability to secrete insulin, and
the incorporation of radioactively labelled uridine into total
cellular RNA as a function of glucose concentration in the incubation medium were compared in islets purified by hand or by
Metrizamide gradients. In respect of admixture with exocrine
tissue, as was confirmed by phase-contrast microscopy, islets
purified on Metrizamide gradients were purer than those purified
by hand as judged by the four times higher α-amylase specific
activity of the latter (Table 1). The glucose-mediated

stimulation of insulin secretion is preserved in islets prepared on Metrizamide gradients (Fig. 1), although the absolute amount of insulin secreted is slightly decreased. This is a distinct improvement over former reports on islet preparation in gradients. Islets purified on sucrose or Ficoll gradients showed a lack of uniform response and little or no

Table 1. Assays for α-amylase as a measure of islet purity.

Tissue material	α-Amylase s.a., units/ mg prot.	% of homogenate activity
Pancreas homogenate	44.0	(100.0)
Islets purified by hand	2.7	6.1
Islets purified with Metrizamide gradients	0.7	1.6

stimulation by glucose. Glucose-mediated stimulation of uridine incorporation into total cellular RNA is evident in both islet preparations (Fig. 2). The overall specific radioactivity was one-third to one-half lower in different Metrizamide-treated preparations. This may be caused by a reduced uptake of [^{3}H]-uridine into the cells.

When Lernmark et al. [6] purified islets by use of a gradient of Ficoll, which was dialyzed beforehand to avoid a toxic effect on cells (personal communication), glucose gave a several-fold stimulation of the synthesis and the release of insulin. Chick et al. [7] also used dialyzed Ficoll for purification of intact pancreatic islets. They lost some of the islets in the sediment. This loss is nearly avoided in Metrizamide gradients. The gradient medium Percoll was used by Buitrago et al. [8] for the isolation of islets from collagenase-digested pancreas by sedimentation at unit gravity. The glucose-mediated stimulation of insulin release was slightly decreased in islets purified through Percoll (as for Metrizamide) compared with by hand.

LITERATURE RELEVANT TO POSSIBLE SEPARATION OF INDIVIDUAL CELL TYPES FROM ISLETS

To meet a request from the Editor, some literature searching has been undertaken. Work is now being done by T. Andersson in collaboration with H. Pertoft (cf. #B-4, this vol.) to separate alpha and beta cells from purified islets. Generally, however, reports on separation attempts seem to be lacking. Probably the small amount of tissue one gets after purifying islets has hitherto precluded further fractionation of their cells. A way out of this shortage of material has already been considered above . Additional promising ways to get access to enough starting material are the cultivation of islets in vitro and the induction of tumours of one cell type.

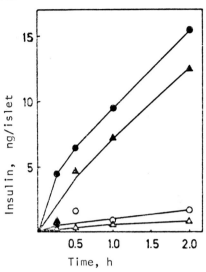

Fig. 1. Glucose-dependent stimulation of insulin secretion, measured by RIA. Islets collected by hand (o, ●) or purified on Metrizamide gradients (△, ▲), incubated with 5 m-mol/1 (o, △), or with 15 m-mol/1 (●, ▲) glucose.

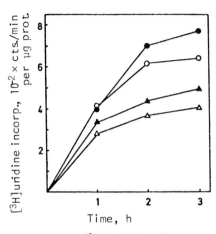

Fig. 2. [³H]Uridine incorporation into acid-precipitable material from islets, expressed on a protein basis, in relation to glucose concentration in the medium. Symbols as in Fig.1.

Spontaneous alpha- or beta-cell adenomata or carcinomata may arise in man. Fischer et al. [9] reported on experiments to establish permanent cell lines from human beta-cell tumours. Experimentally induced tumours of beta cells were produced by Rakieten et al. [10] by the combined action of streptozotocin and nicotinamide, and by Chick et al. [11] following high-dose whole-body X-irradiation of rats each in parabiosis with another rat protected from radiation. He obtained transplantable tumours secreting insulin. These were used by Gilbert [12] for purifying the insulin m-RNA and preparing c-DNA on it as a template. The double-stranded c-DNA was integrated into a penicillinase gene on a plasmid and put into E. coli, which synthesized small amounts of insulin coupled to penicillinase. A good review of the problems connected with pancreatic beta-cell culture has appeared [13]. Generally the culture of islets was difficult for years because of slow growth of the cultured cells over short periods of time. With the exception of the guinea pig [14], the pancreatic tissue was taken from embryos or newborn animals [15]. Recently Ohgawara et al. [16] managed to culture islets from normal adult rats in monolayer in the presence of the phosphodiesterase inhibitor 3-isobutyl-1-methylxanthine and high glucose.

A promising method is the culture of islets in hollow fibres [17]. These allow a continuous nutrient supply to and waste removal from the cells located in an extracapillary space. Cell lines and primary neoplastic cells of several kinds, not pancreatic initially, were maintained and grown to solid-tissue density in this *in vitro* system as developed by Knazek *et al.* [18]. The walls of the fibres allow permeation of small molecular secreted hormones, but are essentially impermeable to antibodies and lymphocytes. Cultured islets are therefore protected from immune rejection; hence such fibres filled with islets as an artificial pancreas came to be implanted into experimental diabetic rats [19, 20]. In an attempt to obtain alpha-cells enrichment in an islet monolayer culture Nakhooda *et al.* [21] directly added to established culture dishes the beta-cytotoxic agent streptozotocin, or treated newborn rats with streptozotocin prior to use of their pancreases in culture. They found an almost complete absence of insulin in either the culture medium or extracts of cells. Glucagon release was unaffected. The glucagon contents of the cells, however, showed a significant decline as compared to controls, possibly as a direct toxic effect of the drug. This result, however, could otherwise imply that after successful preparation of a single cell type from islets, che experiments on it may be affected by a hormone deficiency, insofar as the hormone-secreting cells normally near the cell type under examination have been separated out.

References

1. Lacy, P.E. & Kostianovsky, M. (1967) *Diabetes 16,* 35-39.
2. Gerner, R., L'Age-Stehr, J., Den Tjide, T. & Wacker, A. (1971) *Hoppe-Seyler's Z. Physiol. Chem. 351,* 309-311.
3. Helmke, K., Slìjepčević, M. & Federlin, K. (1975) *Horm. Metab. Res. 7,* 210-214.
4. Loos, J.A. & Roos, D. (1976), Seglen, P.O. (1976) & Freshney, R.I. (1976) in *Biological Separations in Iodinated Density Gradient Media* (Rickwood, D., ed.), Information Retrieval, London, pp. 97-105, 107-121, and 123-130 respectively.
5. Raydt, G. (1977) *Hoppe-Seyler's Z. Physiol. Chem. 358,* 1369-1373.
6. Lernmark, Å., Nathans, A. & Steiner, D.F. (1976) *J. Cell Biol. 71,* 606-623.
7. Chick, W.L., Like, A.A. & Lauris, V. (1975) *Endocrinology 96,* 637-643.
8. Buitrago, A., Gylfe, E., Henriksson, C. & Pertoft, H. (1977) *Biochem. Biophys. Res. Commun. 79,* 823-828.
9. Fischer, E.J., von Kalinowski, H. & Fussgänger, R. (1977) *Diabetologia 13,* 392.
10. Rakieten, N., Gordon, B.S., Beaty, A., Cooney, D.A., Davis, R.D. & Schein, P.S. (1971) *Proc. Soc. Exp. Biol. Med. 137,* 280-283.

11. Chick, W.L., Warren, S., Chute, R.N. & Lauris, V. (1976) *Diabetes 25,* 344.
12. Gilbert, W. (1978) *Cited in Nature (Lond.) 273,* 485.
13. von Wasielewski, E. & Chick, W.L. (editors) (1977) *Pancreatic Beta-cell Culture* (Hoechst Workshop Conferences; Vol. 5; Internat. Congress Series No. 408), Excerpta Medica, Amsterdam.
14. Moskalewski, S. (1965) *Gen. Comp. Endocrinol. 5,* 342-353.
15. Lambert, A.E., Blondel, B., Kanazawa, Y., Orci, L. & Renold, A.E. (1972) *Endocrinology 90,* 239-248.
16. Ohgawara, H., Carroll, R., Hofmann, C., Takahashi, C., Kikuchi, M., Labrecque, A., Hirata, Y. & Steiner, D.F. (1978) *Proc. Natl. Acad. Sci. U.S.A. 75,* 1897-1900.
17. Chick, W.L., Like, A.A. & Lauris, V. (1975) *Science 187,* 847-849.
18. Knazek, R.A., Gullino, P.M., Kohler, P.O. & Dedrick, R.L. (1972) *Science 178,* 65-67.
19. Tze, W.J., Wong, F.C., Chen, L.M. & o'Young, S. (1976) *Nature (Lond.) 264,* 466-467.
20. Chick, W.L., Perna, J.J., Lauris, V., Low, D., Galletti, P.M., Panol, G., Whittemore, A.D., Like, A.A., Colton, C.K. & Lysaght, M.J. (1977) *Science 197,* 780-782.
21. Nakhooda, A.F., Wollheim, C.B., Blondel, B. & Marliss, E.B. (1977) in *Pancreatic Beta-cell Culture,* as for ref. 13, pp. 119-133.

APPENDIX:
Outline of the
procedure [5]
cited on p. 138.

ISLET PREPARATION

Collagenase digestion: P.E. Lacy and M. Kostianovsky

3 x sedimentation at 1 x g in Hanks albumin solution

collection of islets
under the stereomicro-
scope by hand

metrizamide gradient

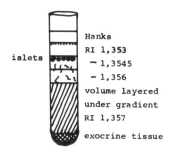

islets

Hanks
RI 1,353
– 1,3545
– 1,356
volume layered
under gradient
RI 1,357
exocrine tissue

1000xg, 5', 4°C, swing out centr.
buoyant density of islets: 1,354

#C-5

IDENTIFICATION, SEPARATION AND CULTURE OF MAMMALIAN TISSUE CELLS

ROBERT J. HAY
Cell Culture Department,
American Type Culture Collection,
Rockville, Maryland, U.S.A.

Source and type of cell *Post-natal guinea pig (200-300 g), monkey, adult; human, various ages. Cell types include pancreatic (acinar) and lung (various).*

Dissociation *Collagenase (2.5-5.0 mg/ml) and trypsin (1.2-2.0 mg/ml) in Hanks' saline without divalent cations, pH 7.4-7.6: gyrate at 120 rev/min at 37° for 15 min periods (up to 4). Collect dissociated cell populations in medium after filtration through nylon mesh (20-40 μm pores).*

Separation *Pancreatic acinar cells recoverable after three sequential sedimentations (5 min each) at 100 g through 4% bovine serum albumin (BSA) in saline. Lung cells separable by clonal culture techniques with selection by altering inert substrates or by pre-treatment with serum exhibiting anti-lung-fibroblast activity.*

Products *Acinar epithelia: up to 95% purity as determined by count of cells with zymogen inclusions; yields consistently ∿30% from the initial dispersed cell populations; viability 85-95% by dye exclusion.*

Lung cells: ∿2-3 × 10⁴ viable cells/mg tissue, ∿90% viable; clone-forming efficiencies varied with donor, species and conditions chosen but were generally 0.3-4% in primary cultures. Fibroblast-like colonies and presumptive type 2, Clara, endothelial and goblet cell clones were located, with identification relying essentially on morphological and ultrastructural examinations. Extensive propagation under the conditions employed was confined to fibroblast-like and one clonally-derived, presumptive type II cell strains.

Comments   *Three general approaches have been utilized to obtain*
            specific cell types from the pancreas or lung in
            culture. —
            (1) Quantitative clonal-cell culture, which serves
            both to isolate specific cell types and to define
            appropriate culture variables.
            (2) Aggregation of specific fractionated epithelia,
            with or without application of selection methods, and
            subsequent culture and study of the cells isolated as
            colonial aggregates.
            (3) Co-cultivation schemes with irradiated feeder
            layers to support survival and division of fractionated
            and characterized tissue cells.

The above-mentioned three methodological approaches for the isolation and culture of specific cell types are now amplified with
particular reference to work with pancreas and lung. Ideally,
cells with readily identifiable markers should be followed from
their source tissue, through appropriate dissociation and fractionation procedures, then during subsequent maintenance *in
vitro*. In cases where this sequential study is possible the
investigator can be certain of cellular identities even if the
specific marker changes or is lost during incubation.

MATERIALS AND METHODS

Tissue sources

Donor species for the work to be described include humans of
various ages, adult Rhesus monkeys, and young guinea pigs (200-
300 g). The latter were either Strain 13 (inbred) or Hartley
strain animals.

Dispersion

The dissociation and fractionation methods used for pancreatic
acinar tissue have been described in detail elsewhere [1-3].
The technique employed for lung tissue, outlined in Scheme 1,
was developed after an extensive series of trials to identify
suitable procedures. Criteria for evaluation of the various
methods tested included yields of dye-excluding cells per unit
weight of starting tissue and clone-forming efficiencies of
cells from recovered, mixed populations.

Clonal cell culture

Standard methods were used throughout [4]. Unless otherwise
specified, F12K [5] supplemented with 15% (v/v) selected foetal
bovine serum (F12K15) was used as the basic medium. Inoculation densities were varied from 500 to 5×10^6 per 9 cm plate
depending on the experimental evaluations required. Plates
coated with collagen gel were prepared as described by Ehrmann &
Gey [6]. Plates with fixed cellular substrate were prepared by

treating confluent monolayers of fibroblasts for 2-3 h with
phosphate-buffered glutaraldehyde (0.1 M, pH 7.4, 2%). Subsequently these were rinsed every 8-16 h for a 4-5 day period with
sterile Hanks' saline (HBSS) and were stored at 4° until use.
The BSA polymer was prepared as described by Macieira-Coelho et
al. [7].

Feeder layers

Feeder-cell populations of lung fibroblasts (ATCC.CCL 171) were
prepared by irradiating cell pellets with 4000-6000 rads using
either an X-Ray machine (Depth therapy generator, Phillips RT250)
or gamma irradiator (Gammator B, Isomedix Inc., New Jersey).
Cultures having accrued 25-30 population doublings (PDL) were
used. Recipient plates were seeded with 1.6×10^4 cells/cm^2 in an
appropriate volume of medium (e.g. 3 ml/T25) to yield cultures
of about one-third confluency. Fluids were renewed every 2-3
days during each experiment.

Electron microscopy

Clones on polystyrene dishes were processed for e.m. examination
as described by Rash & Fambrough [8]. Cell pellets were processed similarly but were infiltrated and embedded in Epon
essentially as described by Luft [9].

Immunological selection

(a) Cytotoxicity assays

A standard assay for antifibroblast antibodies in test sera was
performed as follows. — Sera in replicate tube wells were diluted
using F 12K 15 buffered with HEPES (10 mM). Test cells at 10^6/ml,
likewise in HEPES-buffered F 12K 15, were added in volumes equal
to those of the diluted serum. After incubation for 1 h at 37°
with constant agitation (40 rev/min) the cells were collected by
centrifugation (100 g for 5 min) and were re-suspended in HBSS
containing erythrocin B (0.4 g/100 ml). The tubes or dilution
plates were placed on ice and titres were determined by direct
viable cell counts to identify the 50% endpoint.

(b) Preparation of antibody

Antibody to a clonally-derived strain of guinea-pig-lung fibroblasts (JH$_4$, Clone 1, ATCC-CCL 158) was prepared in female blue-
Dutch rabbits. Pre-immune sera were negative at a 1 : 10 dilution
when tested for anti-fibroblast activity. Fibroblasts at passages 4-8 (20-35 PDL) were recovered from monolayers by trypsinization, and were washed and re-suspended in HBSS minus divalent
cations prior to use. Injections ($1-2 \times 10^7$ cells/dose) were
spaced two weeks apart, and antisera titres were monitored using
serum samples collected on alternate weeks. The first two injections were administered intravenously. The third injection was

Scheme 1. Procedure for dissociation of lung tissue.

Kill animal by decapitation and exsanguinate. Proceed *rapidly* as follows. —

<u>1</u>. Shave thorax, sides, neck and forelimbs.

<u>2</u>. Swab the entire carcass liberally with 70% ethanol. *Asceptic precautions are employed with all subsequent manipulations.*

<u>3</u>. Make an incision along the ventral midline, exposing the pleural and pericardial cavities. Perfuse the lungs with 30 ml HBSS without divalent cations, using an 18 G needle and a 30 ml syringe. Administer the perfusate *via* the right ventricle after opening the left atrium to reduce outflow resistance.

<u>4</u>. Prepare the mixture(s) of trypsin (standard 1.9 mg/ml) and collagenase (standard 2.5 mg/ml) just prior to use from working stocks (trypsin, 0.25%; collagenase 10 mg/ml; 1800 U/ml). Generally, 50 ml is used per 2-3 g lung (assuming guinea pig tissue is from a 250 g animal). Inject 10 ml *via* the right ventricle as in Step 3.

<u>5</u>. Remove the lungs and trim free of the primary bronchus and extraneous tissue and membranes. Transfer to a tared petri plate and weigh.

<u>6</u>. Mince thoroughly with lens scissors such that the size of the lung fragments is ~2-3 mm³. Transfer the fragments using a 'spoonula' to a 50 ml Erlenmeyer flask.

<u>7</u>. Add 20 ml of dissociation fluid, washing the petri plate with the final aliquot. Wide-bore pipettes may be used to advantage at this stage. *This is the initial step for pre-dissected human or monkey lung tissue.*

<u>8</u>. Dissociate with agitation (140 oscillations/min) in a shaking water bath at 37°. The time chosen varies with donor age, being about 15 min/step with embryonic and 30 min/step with older animals.

<u>9</u>. Again using a wide-bore pipette, remove the supernatant and strain through about 6 layers of cheesecloth fitted to a Buchner funnel. Wash residue in the cloth by adding 10-20 ml F 12K 15. Wring out the cheesecloth by gently rolling it using the sterile pipette.

<u>10</u>. Repeat steps 7-9 (one repeat is usually sufficient to disperse most fragments if the original mincing was adequate).

<u>11</u>. Pool the filtered cell suspensions and pass the fluid through Buchner funnels fitted with 'nitex', 40 μm and 20 μm pore sizes. Wash with 10-20 ml F 12K 15. ('Nitex' is nylon mesh.)

<u>12</u>. Separate the suspended cells by centrifugation at 100 *g* for 15 min. If the supernatant is extremely viscous it can be

diluted using F 12K 15 containing hyaluronidase (0.4%) and/or DNase (300 µg/ml) to reduce viscosity and facilitate sedimentation of dispersed lung cells.

<u>13.</u> Discard supernatant and re-suspend the cell pellet in F 12K 15.

<u>14.</u> Dilute an aliquot in a saline solution of erythrocin B and perform routine cell counts. Dilute the original suspension and inoculate into culture vessels as appropriate.

---

also given i.v., but 1 ml Benedril (10 mg/ml, Parke-Davis) was incorporated with it to prevent anaphylactic shock. Each successive injection was administered subcutaneously with 1 ml Freund's complete adjuvant (Perrin's modification, Calbiochem) added to the cell suspension to enhance antibody production. Injections were continued for 6-7 months or until a titre >1 : 500 was achieved.

(c) Immunological selection of cells

Species-specific antibodies were absorbed from the rabbit serum by mixing with cultured primary cells from guinea pig liver. The species-specific fraction of the antiserum was considered removed when 50% of the liver cells were viable after a 1 h incubation period in the antiserum. After such absorption, the antiserum was diluted 1 : 10 in HEPES-buffered F 12K 15 and was sterilized by Millipore filtration. The fluid was then added to a suspension of freshly dissociated guinea pig lung cells to give a final concentration of 5×10^6 cells/ml in a 1 : 20 dilution of serum. The cell suspension was incubated in a water bath shaker at 37° for 1 h at a constant agitation of 40 oscillations/min. The viability of the cells was determined, and the cells were seeded into 9 cm petri dishes at various inoculation densities.

EXOCRINE PANCREATIC CELLS OBTAINED

Clonal analyses permitted definition of culture variables and isolation of epithelial-like clones from human and guinea pig tissues. However, zymogen inclusion droplets and amylase were not retained within such cell groupings after 1-2 weeks *in vitro*, and there was a lack of extensive cell proliferation [1, 2]. Accordingly, we proceeded to the second methological approach. Acinar cells, identified by zymogen inclusions, were recovered after fractionation from primary cell populations of the dissociated tissue [1-3]. After aggregation and plating these were found to attach and form aggregate colonies on standard polystyrene culture dishes (Plate 1). Although zymogen droplets disappeared within one week, the identity of the cells under study was established since the population used for initial seeding contained up to 95% exocrine cells. Furthermore, cells

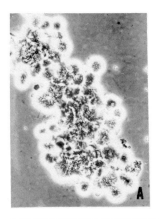
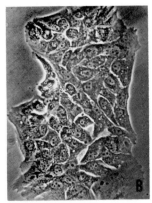
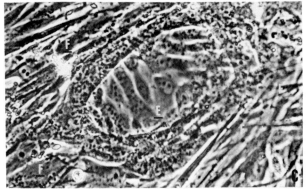

Plate 1. Phase-contrast photomicrographs.
(A, *i.e. top left*) Aggregate of pancreatic acinar cells at time of seeding, x 170;
(B) colonial aggregate after 4 days in culture, x 170;
(C) acinar epithelia (E) after two passages and 5 weeks in co-culture with irradiated human-lung fibroblasts (F), x 375.

with zymogen inclusions could be observed directly as they attached to the plastic substrate. These populations likewise did not exhibit extensive proliferation but could be maintained for 3-4 weeks *in vitro* before degenerating [1].

More recent work involved the third approach, viz. co-cultivation of such cell aggregates with irradiated 'feeder layers' of human fibroblasts. These studies have shown that pancreatic epithelia are much more active in co-cultures of this kind as demonstrated by cell count, by morphological evaluations (Plate 1, C) and by incorporation of tritiated thymidine [10].

LUNG CELLS OBTAINED

i) Dispersion

Variables relating to dissociation of this tissue were evaluated both in terms of viable cell yields and by assessing clone-forming abilities of resultant cell populations. Table 1 summarizes the results of one series of experiments in which the

Table 1. Effects of various combinations of trypsin and collagenase on yields of viable cells obtained on dissociation of
guinea pig lung.
Dissociation was done with pre-warmed solutions (37°; *see text*).
T = trypsin, 1 : 250 (Difco), C = crude collagenase (Type I *ex*
Sigma); stock solutions (10 mg/ml, 1800 U/ml) were dialyzed and
centrifuged prior to use [4]. The values represent mg/ml in
saline buffer at pH 7.6. Viability was assessed by the dye-
exclusion test using Erythrocin B.

concentrations of tryp-
sin and collagenase
were varied. Note that
trypsin at 1.9 mg/ml
and collagenase at
2.5 mg/ml were gene-
rally most satisfactory.
These concentrations
were employed in sub-
sequent experiments.
The fine-structural
morphology of cells
recovered after such
treatment is depic-
ted in Plates 2 & 3.
Pulmonary cell types
I and II, macrophages,
fibroblasts, goblet
cells, ciliated cells
and endothelial-like
cells were observed.

Expt. no.	*Disso-ciation mixture* T	C	Total viable cell yield ×10⁻⁶	Viabi-lity, %	Viable yield ×10⁻³/mg tissue
L5	1.25	5	13	76	18
	2	1.4	4.3	76	6.7
	1	0.7	2.9	78	6.4
	0.5	0.3	0.4	57	0.7
L20	2.5	-	2.6	89	7.8
	1.9	2.5	4.4	90	12
	1.2	5	11	94	18
	0.6	7.5	0.4	80	1.8
L21	1.9	2.5	7.6	88	6.9
	1.25	5	24	87	18
L22	1.9	2.5	25	92	18
	1.25	5	27	87	17
L23	1.9	2.5	36	86	21
	1.25	5	26	83	17
L26	1.9	2.5	67	88	31
L27	1.9	2.5	35	92	34

Aliquots from such
populations which were
seeded for establish-
ment of mass cultures
retained cell types
with characteristic
inclusions over the short term (e.g. type 2, Plate 4). However,
within a few passages the predominant cell type had fibroblas-
tic morphology. This overgrowth by fibroblast-like cells is a
common problem in cell-culture studies that require observations on specific, non-fibroblastic cell types. Cultures were
therefore established using inoculation densities and conditions
which would yield clones.

ii) Selection by substrate alteration

The average total clone-forming efficiencies varied somewhat
among replicate experiments with alterations in substrates and

Plate 2. Elec-
tron micro-
graphs of
guinea-pig
lung cells
after dis-
sociation as
shown in
Scheme 1.
Note presence
of types 1
and 2 cells
as well as
macrophages
(M) and fibro-
blasts (F).
x 5300.

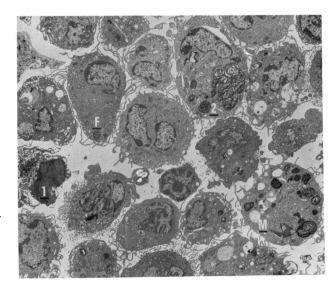

Plate 3.
Electron
micrographs
of guinea-
pig lung
cells after
dissociation.
(A) Note
ciliated
(C) and
goblet (G)
or secre-
tory cells
with mucous
droplets.
x 1700.

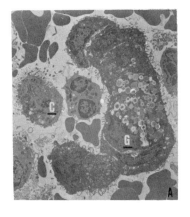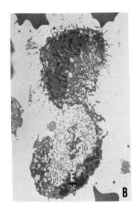

(B) Endothelial-like cells (E).   x 2900.

media, and between species.   Table 2a summarizes results from
two experiments with populations from adult monkey lung.   Note
that cultures on standard polystyrene petri plates (Falcon 3003)
consisted predominantly of clones exhibiting fibroblast-like
morphology. Epithelial and endothelial-like clones were located with much lower frequency. In general, this was a consistent finding with lung-cell populations from all species tested.
Similar data with populations from adult human lungs are summarized in Table 2b.

    The high developmental frequency of fibroblast-like clones

Plate 4.
Electron
micrograph
of human
lung cells
after 48 h
in culture.
Note pre-
sence of
characte-
ristic
osmiophilic
lamellar
inclusions,
x 6800.

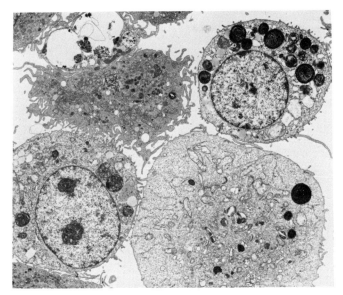

prompted attempts to delineate methods for selection of epithelial and/or endothelial-like colonies. The use of collagen gel
as a substrate proved to be one suitable technique for this purpose. Clonal analyses shown in Tables 2a and 3a clearly suggested that development of epithelial and endothelial-like clones
were favoured when collagen gel was employed as substrate.
Unfortunately, the selective effect was more evident at higher
inoculation densities. Therefore, physical isolation of epithelial cell populations under such conditions was more difficult
(Plate 5).

Use of other inert substrates (Table 3b) offered no apparent
advantage over the polystyrene control for selection of epithelial-like clones. However, endothelial-like colonies were observed in higher ratio on the glass substrate as opposed to
polystyrene controls.

iii) Selection after treatment with cytotoxic antiserum

Work from other laboratories [11, 12] suggested that antisera to
fibroblasts might be made sufficiently specific to permit cytotoxic discrimination against this cell type. Table 4 provides
analytic data from an experiment designed to test this conten-
tion using primary cell suspensions from guinea-pig lung. It is
apparent that pre-absorbed, anti-fibroblast serum effectively
selected against fibroblasts. Note that a partially cytotoxic
action against epithelia was also evident. A three-log increase
in inoculation density did not yield a proportionate increase in
epithelial clone-forming efficiency.

Table 2. Clonal analyses on plates seeded with cells from (a)
Rhesus monkey lung (adult), (b) human lung.
For weighing and dissociation of lung tissue, *see text*. In (a), viability was 87-90%, and yield ∿1.5 × 10⁴ cells/mg tissue. Gross morphology denoted F = fibroblast-like, EN = endothelial-like, E = epithelial-like.

Expt. no.	Passage	Inocula-tion dens., cells/9 cm plate	Substrate	Av. no. clones /9 cm plate vs. morphology			Av. total clone-forming effici-ency (C.F.E.), %
				F	EN	E	
(a) *Monkey lung*							
L87	1	10⁴	Polystyrene	13	2	1	0.16
	2	400	{ Polystyrene	17	0	1	4.5
			{ Collagen gel	8	1	20	7.2
L78	1	10⁴	Polystyrene	23	-	2	0.25
(b) *Human lung*							
H95*	1	10⁴	Polystyrene	21	8	4	0.33
H96	1	10⁴	Polystyrene	32	2	2	0.36
H97	1	10⁵	Polystyrene	16	7	29	0.05
	24-S†	10⁵	Polystyrene	27	3	50	0.08
H98	1	10⁴	Polystyrene	20	1	6	0.27

*H95: *female, age 44; chronic inflammation of left upper lobe;
fibrosis and bronchial dilation. Benign carcinoid tumour removed at margin of resection with bronchus of left upper lobe.
Normal tissue used for cell culture.*
H96: *white female, aged 45; benign cartilaginous haematoma in
right lung. Normal tissue surrounding lesion used for culture.*
H97: *white male, age 49; acute, chronic and organizing pneumo-
nitis with abscess formation and pulmonary fibrosis. Normal
tissue used for culture.*
H98: *white female, age 45; granulomatous lesion. Normal tissue
from biopsy and around lesion used for culture.*
†*Cells adhering after 24 h re-dissociated and used for analysis
and storage.*

iv) Clonal isolation and initial characterizations

The methods described yielded clones of differing morphologies
which could be isolated and examined for characteristic inclusions. In some cases it was possible to propagate progeny from
such isolates to generate cell populations which could be frozen,
characterized and retained for future study.

Cells of one such line (6D1) were found by e.m. to contain
osmiophilic lamellar bodies (Plate 6). However, the number of

Table 3. Clonal analyses on plates seeded with guinea-pig lung cells: (a) effect of collagen gel, *vs*. polystyrene as control, and of varying inoculation density; (b) effect of certain substrates *vs*. polystyrene as control, including polymerized serum albumin ('P-SA') — prepared as in ref. [7]. For F, EN, E *see* Table 2.

Subs-trate	Inoc-ul'n dens-ity	Av. no. clones /9 cm plate *vs.* morphology			C.F.E. %	% E
		F	EN	E		
(a) *Influence of collagen gel & of inoculation density*						
CONT.	10^3	9	15	1	2.5	4
	10^4	63	41	1	1.1	1
Colla-	10^3	6	19	2	2.7	7
gen	2×10^3	10	30	5	2.2	11
gel	10^4	82	142	80*	3.0	26
CONT.	10^3	21	12	4	3.7	11
	3×10^3	49	37	7	3.1	8
	6×10^3	111	55	27	3.2	14
Colla-	10^3	14	8	2	2.4	8
gen	3×10^3	37	29	15	2.7	19
gel	6×10^3	87	96	69*	4.2	27
CONT.	10^3	19	8	2	2.9	7
	10^4	144	107	44	3.0	15
Colla-	10^3	22	2	6	3.0	20
gen gel	10^4	92	47	79*	2.2	36
(b) *Other substrate influences*						
CONT.	10^3	9	6	2	1.7	13
F F mat†	10^5	60	25	0	0.08	-
Glass	10^3	15	24	1	4.0	3
CONT.	10^3	32	13	5	5	10
P-SA	10^5	27$^\nabla$	2	0	0.03	-

Epithelial clones often present as large patches overlapping adjacent colonies.

†*Fixed monolayers prepared as described in the text; guinea pig lung fibroblasts in the fourth passage were utilized.*

∇ *Clones small - only 8-32 cells.*

cells in the population retaining these inclusions was <10%. Furthermore, biochemical studies did not reveal high levels of phosphatidylcholine synthesis [13] when comparisons were made with control fibroblast cultures. One would expect a high synthetic rate if the cells were of type II origin and retained specific functional activity under the culture conditions imposed.

The ideal of obtaining continuously-propagable epithelial lines from isolated colonies was attained very infrequently. In most cases, isolated epithelial populations (Plate 7) showed reduced proliferative activity and degenerated within 1-3 passages. Initial characterization work was therefore directed towards attempts to identify clones with distinctive morphologies by examination of cell fine structure. To accomplish this, the clonal cultures were fixed and embedded *in situ* [8], then sectioned tangentially to the culture surface for

Table 4. Fibroblast destruction by the prepared antiserum, shown by clonal analysis.
Guinea pig tissue was dissociated to provide the test cell suspension. For antiserum preparation, *see text*. F, EN and E are defined in the heading to Table 2; values for duplicate plates are given. As before, inoculation density = cells/9 cm Petri plate.

Treatment	Inoc-ul'n dens.	No. of clones/plate *vs.* gross morphology			Total	% fibro-blast
		F	EN	E		
Incub. 1 h in HEPES-buffered F 12 K 15	10^3	45, 34	6, 7	7, 14	113	70
	3×10^3	110, 112	25, 16	9, 26	298	74.5
Incub. 1 h in a 1:10 dilution of antiserum in HEPES-buffered F 12 K 15	5×10^5	0	1	3, 8	12	0
	10^6	0	0	34, 39	73	0

Plate 5. Guinea pig lung cell clones on a polystyrene plate (P) and on a plate coated with collagen gel (G) from rat-tail tendons. Note the large epithelial-like clones which developed on the gel substrate *(arrows)*. Both plates were seeded with 10^4 cells from primary suspensions.

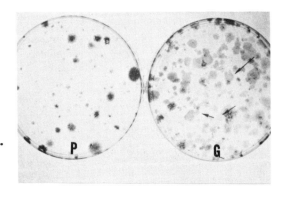

Plate 6. Electron micrographs of a clonally-derived cell strain (6D1) from guinea-pig lung. The cells (A)

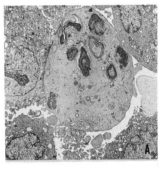
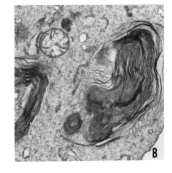

were processed for microscopy after 6 passages (about 23 population doublings). × 1700. Note the lamellar structures shown at higher magnification, × 12000 (B), suggesting that the strain may be of type 2 origin.

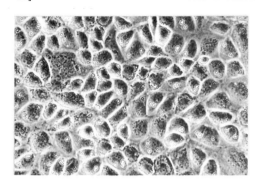

Plate 7. Living cells of
an epithelial-like clone
which develop in standard
primary cultures of
guinea-pig lung. x 255.

Plate 8 *(below)*. Cultured
epithelial cells from
adult monkey lung.
(A) Early clone. x 170.
(B) Confluent clone.
x 170.

(C) Electron
micrograph pre-
pared from mate-
rial similar to
that shown in
(B), x 7800.
Electron-dense
droplets and
vacuoles were
noted in many
of the isolated
cells.

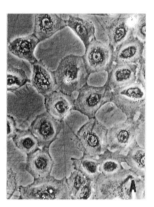
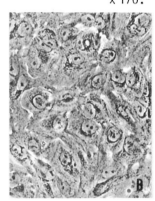

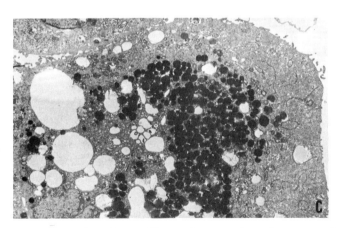

e.m. examination. Photomicrographs of selected clones were prepared prior to fixation to permit correlation of light microscopic (l.m.) morphologies with e.m. images (Plates 8-11).

Evidence for isolation of mucus-containing (goblet) cells is
shown in Plates 8-10. These cell groupings were obtained as

Plate 9. Correlated light and electron micrographs showing appearance of one type of guinea-pig-lung epithelial cell obtained in culture. Note amorphous vesicular cytoplasmic material not unlike that found in goblet cells. x 6800 and x 170.

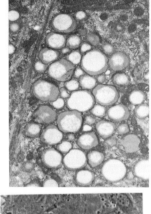
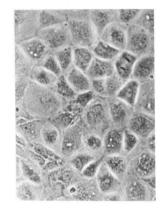

Plate 10. Light and electron micrographs of epithelial cells from guinea-pig lung which developed in culture after treatment with anti-fibroblast antiserum. Note abundance of vesicular droplets. x 170 and x 4100.

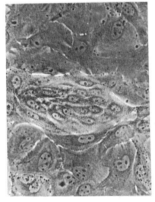
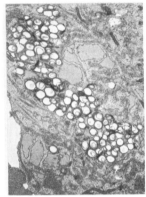

Plate 11. Another correlated sequence. Note dense granules in cytoplasm not unlike some of those seen in Clara cells *in situ*. Guinea-pig lung served as source tissue. x 1700 and x 170.

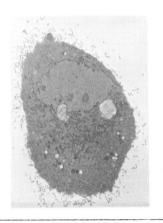
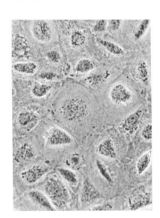

primary clones (Plates 8 & 9) without prior selection or through treatment with anti-fibroblast antiserum (Plate 10). Epithelial-like cells with inclusions not unlike those observed in pulmonary Clara cells are depicted in Plate 11.

Plate 12. Presumptive endothelial cells from guinea-pig lung.
(A, *i.e. top left*) Mass culture plates seeded at $10^5/$ 9 cm. x 210.
(B) Phase-contrast micrograph of a living clone. x 170.
(C) Correlated electron micrograph of clone shown in (B). Note abundance of mitochondria, electron-lucent membrane-bounded droplets and cytoplasmic vesicles. x 6800.

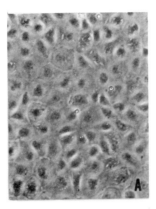
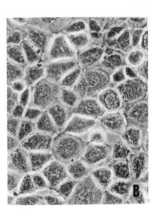
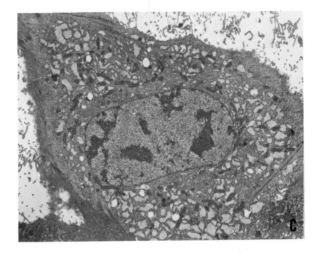

Endothelial-like cells were identified essentially by morphology through comparison with work published elsewhere [14, 15]. These cells were observed in both mass and clonal cultures (Plate 12). Cells from clones selected for study of fine structures did not contain Weibel-Palade bodies; but numerous electron-lucent droplets and cytoplasmic vesicles were present (Plate 12, C).

DISCUSSION

The techniques described have permitted isolation with extended maintenance *in vitro* of specific cell-types from both pancreatic and pulmonary tissues. Each of the three approaches mentioned offers special advantages for work on cell-line development. Cells of the exocrine pancreas can be identified readily by virtue of their zymogen inclusions, prominently visible on examination by light microscopy. Studies with low density

cultures established after inoculation of cell and aggregate
suspensions permitted definition of the conditions whereby
colonial aggregates could be isolated and maintained [1]. Co-
cultivation with irradiated-feeder fibroblasts not only stimulated proliferation markedly but also allowed for extended
maintenance times [10]. Rheinwald & Green [16] reported that
human epidermal keratinocytes could also be propagated extensively in the presence of irradiated feeder-layers. Furthermore the multiplication potential of keratinocytes was markedly
stimulated on addition of epidermal growth factor, giving
populations that exhibited over 150 doublings *in vitro* [17].

The circumstance in terms of cell identification is more
complex when one considers the histological structure of the
lung. This tissue is composed of about 40 different cell
types [18], many of which can most readily be characterized by
examination of tissue ultrastructure [18, 19]. Availability of
cultivable and functional lung lines would not only permit more
reliable definition of their respective roles in pulmonary
physiology but also facilitate investigations of the factors
which regulate their activities. The strictly ideal condition
for study and ultimate development of cell lines, as presented
at the outset, does not hold for lung. With the exception of
the ciliated cells, recognized by their characteristic organelle, viable cells in primary isolates cannot yet be identified. However, a number of immunological, biochemical, histological or cytochemical tests are now available to monitor fractionation procedures.

Thus, isolated pulmonary endothelial cells might be identified by immunological testing for von Willebrande factor VIII
[20] or by biochemical assay for angiotensin-converting activity
[21]. The presence of Weibel-Palade bodies, noted in endothelial cells from human and rat arteries [22], has not been reported for endothelia of the distal lung. Pulmonary macrophages are readily discerned after histo-enzymological staining
for acid phosphatase [23]. Similarly, goblet cells can be recognized after treatment using the periodic-acid-Schiff procedure or by examination of fine structure [24].

Application of one or more of these assay procedures should
permit selection of suitable techniques for separation of some
lung cell species. Reports of success in this regard, for type II
pneumocyte fractionation, were described by Kikkawa & Yoneda [25]
who employed a modified Papanicolaou staining procedure, and by
Mason *et al*. [26] who used the phosphine 3R fluorescent stain to
ascertain reliability of centrifugal separation techniques.
Availability of such methodology for the many different lung-cell
types would enhance our capability for development and study of
lung-cell lines *in vitro*.

The potential utility of clonal culture methods for isolation
of specific pulmonary cells from non-fractionated cell suspensions was explored through studies reported here. The presumptive evidence provided indicates that fibroblast-like cells,
mucus-containing cells, type II cells, Clara cells, and endothelial cells may be recovered as primary clones. The utilization of a collagen gel substrate enhanced clone-forming efficiencies of endothelial and epithelial-like populations. This more
strictly physiological matrix has often been employed successfully for outgrowth of epithelia in explant cultures. A precipitated collagen layer also markedly increases clone formation in
cultures of chick embryonic muscle [27].

Selection against lung fibroblasts was demonstrated using
rabbit anti-lung-fibroblast antiserum pre-absorbed with cultured
liver cells. Surface antigens specific for cultured fibroblasts
have also been noted on populations obtained from human skin [11]
and chick embryonic tissues [12]. The apparent residual cytotoxic action towards epithelia exhibited by absorbed antisera may
reflect a lack of total specificity in the preparations used.
Alternatively, some lung epithelial-cell populations as selected
out may share common antigens with lung fibroblasts. In either
case, the presence of such cell-specific antigens suggests that
immunological selection may provide a valuable supplement for
physical cell-separatory techniques as outlined earlier.

Except for fibroblast-like strains there are very few lung-
cell lines available for study (R.J. Hay, *later in this vol.*).
The rat L2 line (ATCC-CCL 149)   is   one   purported type 2
population available from normal tissue. Although early-passage
L2 cells were found to contain lamellar bodies [28], these
characteristic inclusions were not observed in later-passage
stocks available for distribution (R.J. Hay, *later in this vol.*).
Other lines which express biochemical features of type 2 alveolar
cells were derived from a human lung carcinoma (A549-ATCC-CCL 185)
[29] and mouse lung adenoma (LA$_4$-ATCC-CCL 196) [30].

Once could speculate that difficulties in propagating some
normal lung epithelia may be overcome, at least in part, through
co-cultivation schemes as described for pancreatic acinar cells
from human keratinocytes. The definition of suitable feeder-
layer culture systems may be the subject of future reports.

Acknowledgements

The author acknowledges the participation of J. Hoying, M.S. in
immunological studies and the cooperation of R. Sneider, M.D. in
obtaining human and monkey lung tissues. The expert technical
assistance of B. Arner, P. Armstrong, D. Hornack, N.W. Jessop
and K. Rainey with various aspects of this work is also recognized. The work was supported in part by contract NO1-65751
from the NCI, and NO1-HR-6-2914 and NO1-HR-6-2930 from the NHLBI.

References

1. Hay, R.J. (1975) in *Cell Impairment in Aging and Development* (Cristofalo, V.J. & Holeckova, E., eds.), Plenum, New York, pp. 23-39.
2. Hay, R.J. (1975) *Cancer Res. 35,* 2289-2291.
3. Hay, R.J. (1978) *Tiss. Cult. Assoc. Manual 4,* 809-812.
4. Cahn, R.D., Coon, H.G. & Cahn, M.B. (1966) in *Methods in Developmental Biology* (Wilt, F.H. & Wessells, N.K., eds.), Crowell, New York, pp. 493-530.
5. Kaighn, M.E. (1973) in *Tissue Culture Methods and Applications* (Kruse, P.F. Jr. & Patterson, M.K. Jr., eds.), Academic Press, New York, pp. 54-58.
6. Ehrmann, R.L. & Gey, G.O. (1956) *J. Natl. Cancer Inst. 16,* 1375-1390.
7. Macieira-Coelho, A., Berumen, L. & Avrameas, S. (1974) *J. Cell Physiol. 83,* 379-388.
8. Rash, J.E. & Fambrough, D. (1973) *Dev. Biol. 30,* 166-186.
9. Luft, J.H. (1961) *J. Biophys. Biochem. Cytol. 9,* 409-414.
10. Hay, R.J. & Jessop, N.W. (1978) *In Vitro 14,* 364-365.
11. Bachvaroff, R. & Rapaport, F.T. (1972) *J. Immunol. 109,* 1081-1089.
12. Wartiovaara, J., Linder, E., Ruoslahti, E. & Vaheri, A. (1974) *J. Exp. Med. 140,* 1522-1533.
13. Williams, C.D. & Hay, R.J. (1978) *unpublished observations.*
14. Blose, S.H. & Chacko, S. (1975) *Development, Growth & Differentiation 17,* 153-165.
15. Gimbrone, M.A., Cotran, R.S. & Folkman, J. (1974) *J. Cell Biol. 60,* 673-684.
16. Rheinwald, J.G. & Green, J. (1975) *Cell 6,* 331-344.
17. Rheinwald, J.G. & Green, H. (1977) *Nature 265,* 421-424.
18. Sorokin, S.P. (1970) in *Morphology of Experimental Respiratory Carcinogenesis* (Netteshein, P, Hanna, M.G. Jr. & Deatherage, J.W. Jr., eds.), U.S.A.E.C., Oak Ridge, pp. 3-43.
19. Kilburn, K.H. (1974) *Int. Rev. Cytol. 37,* 153-270.
20. Jaffe, E.A., Nachman, R.L., Becker, C.G. & Minick, C.R. (1973) *J. Clin. Invest. 52,* 2745-2756.
21. Ryan, J.W., Smith, U. & Niemeyer, R.S. (1972) *Science 176,* 64-66.
22. Weibel, E.R. & Palade, G.E. (1964) *J. Cell Biol. 23,* 101-112.
23. Sorokin, S.P. & Brain, J.D. (1975) *Anat. Rec. 181,* 581-626.
24. Meyrick, B. & Reid, L. (1970) *J. Anat. 107,* 281-299.
25. Kikkawa, Y. & Yoneda, K. (1974) *Lab. Invest. 30,* 76-84.
26. Mason, R.J., Williams, M.C., Greenleaf, R.D. & Clements, J.A. (1977) *Amer. Rev. Resp. Dis. 115,* 1015-1026.
27. Hauschka, S.D. & Konigsberg, I.R. (1966) *Proc. Natl. Acad. Sci. 55,* 119-126.
28. Douglas, W.H.J. & Kaighn, M.E. (1974) *In Vitro 10,* 230-242.
29. Lieber, M.M., Smith, B., Szakal, A., Nelson-Rees, W. & Todaro, G. (1976) *Int. J. Cancer 17,* 62-70.
30. Stoner, G.D., Kikkawa, Y., Kniazeff, A.J., Miyai, K. & Wagner, R.M. (1975) *Cancer Res. 35,* 2177-2185.

#C-6
ISOLATION OF ADIPOCYTES

J. P. LUZIO
Department of Clinical Biochemistry,
University of Cambridge,
Addenbrooke's Hospital,
Cambridge, U.K.

*Since its introduction by Rodbell in 1964 [1], the preparation
of intact isolated white fat cells (adipocytes) by digesting rat
epididymal fat pads with collagenase has become a widely used
technique, performed by us as follows :—*

Source *Epididymal fat pads from young male rats (120-140 g).*

Dissociation *The fat pads from 8 rats are placed in a 50 ml,
siliconized glass, conical flask containing 10 ml
Krebs-Ringer bicarbonate medium [2] (1.3 mM Ca^{2+},
pH 7.4, 4% w/v bovine serum albumin, BSA), 10 mg
glucose and 10 mg collagenase ($\sim$140 U/mg). The
flask is shaken at 140 rev/min for 30 min in a water-
bath at $37°$.*

Separation *The isolated fat cells float to the surface of the
cell suspension. To speed this process the suspension is centrifuged at 300 g for 30 sec and the
infranatant and pellet discarded. The cell cake is
washed with 3 x 10 ml of Krebs-Ringer bicarbonate
medium, and re-suspended in this medium.*

Product *Yield $\sim$1 g dry wt (2.6 x 10^7 cells).*

Comments *Cell intactness depends on gentle treatment and use
of siliconized glassware or plastic (polypropylene)
throughout. Hormone responsiveness of the isolated
cells can be particularly affected by the source and
purity of collagenase and BSA.*

*Digestion of fat pads with collagenase to give isolated adipo-
cytes has been widely applied to species other than the rat. The
hormonal sensitivity of the isolated cells differs between
species, and there are also differences depending on the age and*

*nutritional state of the animal, and possibly on the site of
adipose tissue chosen [3]. Isolated fat cells have proved a
highly important model for the investigation of hormone action.*

The rat epididymal fat cell preparation [1] has proved particularly useful because of its sensitivity to a wide variety of
hormones, and the original collagenase digestion protocol has
been used with remarkably little modification to isolate fat
cells from the adipose tissue of many other species and from
different sites in the body. The protocol that we currently use
is now amplified. Consideration is given to problems that may
be encountered when isolating fat cells and to criteria that can
be used to assess the cell preparations.

ISOLATION OF RAT EPIDIDYMAL FAT CELLS

Materials *(supplementary points; cf. description above)*

The pads are removed from young Wistar rats given free access to
food and water and killed by decapitation. Both age and nutritional state of the animal may affect the composition and hormonal sensitivity of adipose tissue *(see below)*. For the preparation of hormonally sensitive fat cells, two items need most careful selection, viz collagenase and albumin. We have always used
collagenase CLS ($\sim$ 140 U/mg) from Worthington Biochemical Corp.,
Freehold, N.J., U.S.A. (British Agents: Cambrian Chemicals,
Beddington Farm Road, Croydon) and BSA (Fraction V) from Armour
Pharmaceutical Co., Eastbourne, Sussex, U.K. Fresh batches of
both collagenase and albumin must be tested for their ability to
prepare viable cells. In particular, albumin batches sometimes
contain insulin-like substances producing an isolated cell
preparation which shows high basal glucose uptake and low stimulation of lipolysis by adrenaline. The amount of collagenase required to isolate cells with our protocol has varied from 8 to 15
mg/10 ml, depending on the batch of enzyme.

The requisite siliconizing of glass vessels is carried out
with 'Repelcote' (Hopkin and Williams Ltd., Chadwell Heath,
Essex, U.K.) or 'Siliclad' (Clay Adams, Parsippany, N.J. U.S.A.).

Dispersion and separation of cells *(supplementary points)*

For the medium, which is gassed with O_2 + CO_2 (95 : 5), the BSA
is made up as a 30% w/v solution in Krebs-Ringer buffer and
dialyzed overnight at 4° against a 10-fold excess of the buffer
before dilution to the final concentration. This ensures that
the final albumin-containing medium is at the correct pH and
ionic composition.

The cell suspension, after the shaking at 140 rev/min in a
water bath (Mickle Lab. Eng. Co., Gomshall, Surrey), is transferred to a plastic centrifuge tube, and any remaining fragments

of undigested tissue removed with forceps. The washed cell cake
from the centrifugation is further washed and re-suspended in
the required experimental medium.

CHARACTERIZATION OF THE ISOLATED FAT CELLS

For fat cells isolated by the collagenase digestion method, the
above-mentioned yield [4] represents, on the basis of DNA measurements, a recovery as high as 80% of those in the initial tissue
[5]. Light microscopy reveals that the preparation consists of
isolated cells which are spherical and are characterized by the
presence of a large central lipid droplet surrounded by a thin
rim of cytoplasm, thickened at one point to accommodate the
nucleus and giving the cell in profile its characteristic 'signet-
ring' appearance (Plate 1a). The cells may vary from 10 to
120 μm in diameter, but in the preparation described are mostly
∿50 μm.

Scanning electron microscopy reveals that the cell surface is
relatively smooth with few ruffles (Plate 1b). Transmission
e.m. of isolated fat cells, as reviewed [6], reveals details of
the cytoplasmic structure (Plate 1c). Although only filling a
small proportion of the cell volume the cytoplasm is highly
organized, containing a smooth e.r. system which surrounds the
central lipid droplet, has close association with mitochondria
and cytoplasmic lipid droplets, and approaches, but does not fuse
with the plasma membrane. The smooth e.r. system may have a
function analogous to that of the sarcoplasmic reticulum in
muscle [7].

Estimates of the intracellular water space and ionic content
have been made. The former is low, reflecting the small amount
of cytoplasm relative to total cell volume. We have determined
a value of 4 μl per 100 mg dry wt of cells [8], and similar
figures have been obtained in other laboratories [9]. Fat-cell
ionic content has been determined by radioisotope efflux studies
and atomic absorption spectrometry. Perry & Hales [4] estimated
the intracellular concentrations of K^+, Na^+ and Cl^- to be 146,
19 and 43 m-equiv/l respectively. Measurement of fat cell
calcium is complicated by the ability of isolated cells to rapidly take up calcium from the medium [10]. It is thought that the
intracellular calcium is largely stored in organelles, since if
evenly distributed throughout the intracellular water the concentration would be about 5 mM, very much greater than the expected cytoplasmic concentration which is believed to lie in the
range 0.1-10 μM [11].

CRITERIA OF VIABILITY OF ISOLATED FAT CELLS

Measurement of hormonal sensitivity is a good criterion, albeit
insufficient in some experiments. Adrenaline stimulation of

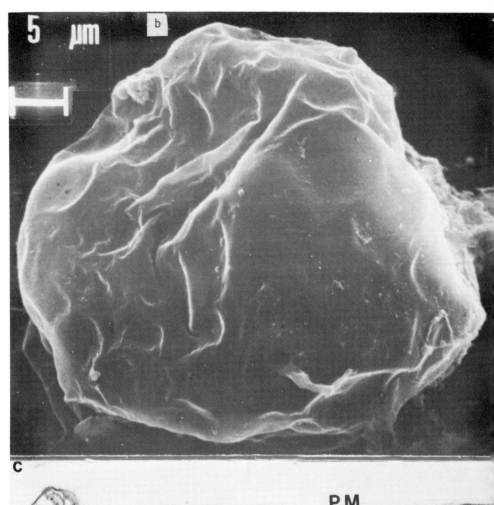

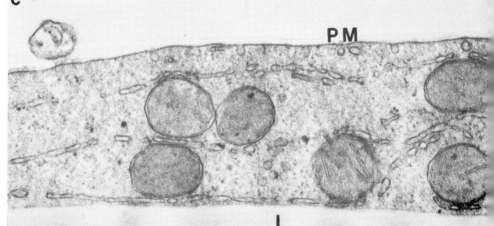

lipolysis, insulin inhibition of adrenaline-stimulated lipolysis
and insulin stimulation of glucose uptake have been widely used
to assess rat isolated fat-cell preparations. The mechanism of
hormonal control of lipolysis has recently been reviewed [3],
and although details are still not clear, there is no doubt that
several hormones stimulate lipolysis and that this is related
to a very rapid rise in intracellular cyclic AMP concentration
(Fig. 1) even when very low concentrations of hormones are used
[3, 12]. Insulin is antilipolytic to submaximal concentrations
of lipolytic hormones, but only when physiological concentrations
are used.

Stimulation of glucose metabolism in rat fat cells by hormones
(in particular insulin) has been measured by different techniques
in different laboratories. The most popular and convenient
methods are conversion of [^{14}C]glucose into [^{14}C]CO$_2$ [1],
labelled glucose into labelled triglyceride [9, 13] and reduction
in extracellular glucose concentration [14]. Although these
measure different aspects of glucose metabolism, hormonal stimulation is rarely greater than 3-5 fold which is very much less
than the >50 fold maximal stimulations of lipolysis that are

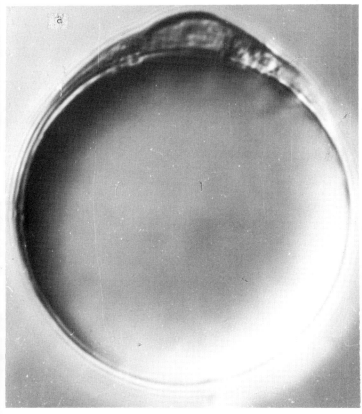

Plate 1.
Micrographs of
rat epididymal
isolated fat
cells.
a: Interference
contrast micro-
graph.
Opposite:
b: Scanning
e.m.
c: Transmission
e.m. (L, lipid
droplet; PM,
plasma membrane),
x 45,000; fixa-
tion and stain-
ing as in [7].

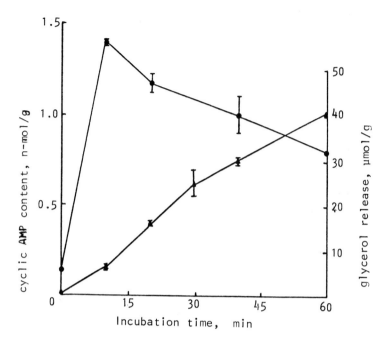

Fig. 1. Time course of stimulation of cyclic AMP content and
lipolysis by adrenaline in rat fat cells. The cells were incubated in Krebs-Ringer bicarbonate medium with 55 μM adrenaline.
Cyclic AMP content (●) and glycerol released (▲) were measured.
Data re-drawn from [3].

often observed. In our experience good stimulation of glucose
uptake by insulin is a particularly useful criterion of cell
viability.

In some experiments other criteria of cell intactness are
required. Both ATP content and release of the cytosol enzyme
lactate dehydrogenase have been used [8] though there are difficulties in interpreting these measurements. In rat fat cells
prepared as described the ATP content, which is 200-400 nmol/g
dry wt. in resting cells, falls greatly in cells whose lipolysis
has been stimulated by adrenaline [15]. Measurement of lactate
dehydrogenase release shows that 10-20% of the enzyme is found
in the medium simply on re-suspending cells, and a further 5-10%
after incubation at 37° for 1 h [8]. It is not known if this
represents leakage or cell breakage; but the latter appears more
likely since membrane enzymes are also found in the medium.

The care required to prepare an experimentally usable isolated
fat-cell population depends to some extent on the type of experi-

Table 1. Glucose uptake in rat epididymal
fat cells isolated in the presence or absence of glucose.
Incubation for 90 min (cell concentration
44 mg dry wt/ml) in Krebs-Ringer bicarbonate buffer containing 2 mM glucose [14],
the final level of which was measured.
Values are expressed as µmoles glucose/g
dry wt cells/h, and are the mean ± SEM of
4 determinations.

	Cells isolated	
	with glucose, 10 mg/ml	without glucose
Basal	2.9 ±1.5	10.8 ±0.9
+ Insulin, 100 µU/ml	17.5 ±0.6	18.5 ±0.6
+ Adrenaline, 55 µM	12.5 ±1.7	14.0 ±1.2

ment to be performed. Lipolysis, for instance, may be affected
by pH, ionic composition of the medium, albumin concentration,
cell concentration and the presence of adenosine [3]. Although
there are reports that glucose can inhibit fatty acid release
due to the stimulation of re-esterification, we have never observed any effect of glucose on the stimulation of lipolysis by
adrenaline in rat isolated fat cells [3, 14].

Investigations of hormonal stimulation of glucose uptake and amino
acid incorporation into fat cell protein are also affected by
the conditions of cell preparation and incubation. Table 1
shows data from a single experiment in which epididymal fat pads
from 12 rats were divided into two portions which were incubated with collagenase in the presence or absence of glucose.
The cells isolated with glucose present subsequently showed a
lower basal rate of glucose uptake than the other cells, but
the hormone-stimulated rate was very similar in both cell preparations. The incorporation of amino acid into fat-cell protein
is also affected by the presence or absence of substrate during
cell isolation. Cells isolated in the absence of amino acids
subsequently showed a distinctly non-linear incorporation of
[³H]leucine into total cell protein, compared with cells isolated in the presence of amino acids (Fig. 2). As with lipolysis the rate of incorporation of [³H]leucine into fat-cell
protein can be affected by cell concentration (Fig. 3).

ISOLATION OF FAT CELLS FROM SPECIES OTHER THAN THE RAT

Isolation of fat cells from adipose tissue pieces of other
species is essentially the same as for rat epididymal fat cells.
However it is often necessary to dice the tissue before incuba-

tion with collagenase and to vigorously shake the incubation
vessel by hand at intervals during the preparation [18, 19]. The
cells isolated often have different characteristics to rat fat
cells.   In particular the sensitivity to hormones of cells from
different species varies widely [3, 20] (Table 2).

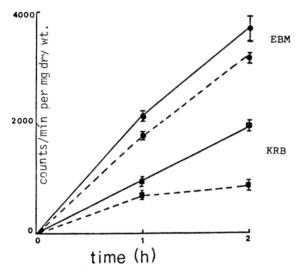

ISOLATED FAT CELLS FROM DIFFERENT SITES

Fig. 2.   Incorpora-
tion of [³H]leucine
into total cell pro-
tein (technique as in
[17]) after isolation
of rat epididymal fat
cells in the presence
or absence of amino
acids.
The cells were incuba-
ted at 70 mg dry wt/ml
in 2 ml Eagle's basal
medium (EBM) [16]con-
taining [³H]leucine
(20 μCi/ml), with
10 μU/ml insulin (——),
or without insulin
(---) in the control.
Cells were isolated
either in EBM (●) or
in Krebs-Ringer bi-
carbonate medium (KRB)
(■) containing 1 mg/ml
glucose.

There is an increasing body of evidence
that the metabolism and hormonal responsiveness of isolated fat cells depends
on the site of the adipose tissue from
which they have been prepared [3, 21; also G. Siebert *et al.*,
Note in this vol.].   This subject is of particular interest when
discussing the deposition and mobilization of triglyceride in
man.   However, many other factors including age, cell size and
diet which can all greatly affect aspects of fat cell metabolism
may be responsible for some observed differences in isolated fat
cells from different sites [3].

CONCLUSIONS

The preparation of isolated fat cells is very simple and is performed in many laboratories.   The rat fat-cell preparation in
particular has proved very useful in the investigation of hormone
action, largely because of the purity of the preparation, its
sensitivity to a wide variety of hormones [22] and the ability
to prepare subcellular fractions [3].   Further investigation of
the cells from other species and from different sites will aid
our understanding of the deposition and mobilization of triglyceride.

Table 2.   Response of lipolysis to
various hormones in different species.
The values (ng/ml) are the minimum dose
to affect lipolysis. N_1 is no effect
detectable. *Data from* [20].

	Adrena-line	ACTH	Insulin	Prosta-glandin E_1
Rat	10	10	0.01	1
Man	1	N	0.004	10
Fowl	1000	1	N	0.1

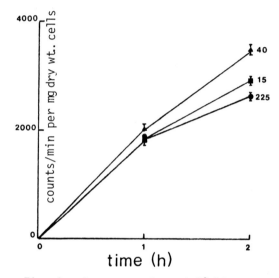

Fig. 3.   Incorporation of [³H]leucine into total
rat fat-cell protein at different cell concentrations.
Fat cells isolated in Eagle's basal medium and
incubated with [³H]leucine as in Fig. 2, but at
cell concentrations of 15, 40 and 225 mg dry
wt/ml.

Acknowledgements

I thank Prof. C.N. Hales for his advice and encouragement,
Prof. L. Herman for the scanning e.m. (Plate 1b), Mr. R. Cohen
for obtaining the data shown in Figs. 2 & 3, and other members
of the Department for their help in many fat-cell experiments.
The work is supported by the Medical Research Council and the
British Diabetic Association.

References

1. Rodbell, M. (1964) *J. Biol. Chem. 239,* 375-380.
2. Cohen, P.P. (1957) in *Manometric Techniques,* 3rd edn. (Umbreit, W.W., Burris, R.H. & Stauffer, J.F., eds.), Burgess, Minneapolis, p. 149.
3. Hales, C.N., Luzio, J.P. & Siddle, K. (1978) *Biochem. Soc. Symp. 43,* 97-135.
4. Perry, M.C. & Hales, C.N. (1969) *Biochem. J. 115,* 865-871.
5. Rodbell, M. (1964) *J. Biol. Chem. 239,* 753-755.
6. Slavin, B.G. (1972) *Int. Rev. Cytol. 33,* 297-334.
7. Hales, C.N., Luzio, J.P., Chandler, J.A. & Herman, L. (1974) *J. Cell Sci. 15,* 1-15.
8. Newby, A.C., Luzio, J.P. & Hales, C.N. (1975) *Biochem. J. 146,* 625-633.
9. Gliemann, J., Osterlind, K., Vinten, J. & Gammeltoft, S. (1972) *Biochim. Biophys. Acta 286,* 1-9.
10. Martin, B.R., Clausen, T. & Gliemann, J. (1975) *Biochem. J. 152,* 121-129.
11. Hales, C.N., Campbell, A.K., Luzio, J.P. & Siddle, K. (1977) *Biochem. Soc. Trans. 5,* 38-44.
12. Siddle, K. & Hales, C.N. (1974) *Biochem. J. 142,* 97-103.
13. Clausen, T., Gliemann, J., Vinten, J. & Kohn, J.G. (1970) *Biochim. Biophys. Acta 211,* 233-243.
14. Luzio, J.P., Jones, R.C., Siddle, K. & Hales, C.N. (1974) *Biochim. Biophys. Acta 362,* 29-36.
15. Stein, J.M. (1975) *Biochem. Pharmacol. 24,* 1659-1662.
16. Eagle, H. (1955) *J. Exp. Med. 102,* 595-600.
17. Miller, L.V. & Biegelman, R.M. (1967) *Endocrinology 81,* 386-389.
18. Burns, T.W. & Hales, C.N. (1966) *Lancet i,* 796-798.
19. Langslow, D.R. & Hales, C.N. (1969) *J. Endocrin. 43,* 285-294.
20. Langslow, D.R. & Hales, C.N. (1971) in *Physiology and Biochemistry of the Domestic Fowl,* Vol. 1 (Bell, D.J. & Freeman, B.M., eds.), Academic Press, London, pp. 521-547.
21. Ashwell, M., Durrant, M., Stalley, S. & Garrow, J.S. (1977) *Proc. Nut. Soc. 36,* 110A.
22. Rodbell, M. (1970) in *Adipose Tissue: Regulation and Metabolic Function [Horm. Metab. Res. Suppl. No. 2],* (Jeanrenaud, B. & Hepp, D., eds.), Georg-Thieme-Verlag, Stuttgart and Academic Press, New York, pp. 1-4.

#D Populations from Non-Tissue Sources

#D-1
HARVEST OF MARINE MICROALGAE BY CENTRIFUGATION IN DENSITY GRADIENTS OF 'PERCOLL'

E. M. REARDON[1], C. A. PRICE[1] and R. R. L. GUILLARD[2]
[1]Waksman Institute of Microbiology, Rutgers University,
Piscataway, N. J., U.S.A.,
and [2]Woods Hole Oceanographic Institution, Woods Hole,
Mass., U.S.A.

Naked dinoflagellates, other fragile flagellates, diatoms and a species of blue-green bacterium have been recovered physiologically active and structurally intact after centrifugation into density gradients of 'Percoll' in a synthetic seawater based on sorbitol. This medium is compatible with Percoll, a polyvinyl pyrrolidone-modified silica sol, over a spectrum of saline and magnesium concentrations typical of those found in situ. *Isopycnic sedimentation of the test organisms into this gradient resulted in the concentration and resolution of the algae in narrow bands, each of characteristic density for the species examined. Both continuous and step gradients were tested, the continuous gradient being linear between a starting density of $\rho = 1.03$ to a terminating density of $\rho = 1.15$. Intactness of the* microorganisms *was judged by comparing microscopic appearance and motility before and after density-gradient centrifugation and after pelleting of the algae. The successful collection of nanoplankton, including some of its most elusive and delicate representatives, has broad implications for the study of the physiology of marine microalgae.*

A salient problem in the isolation of marine microalgae is dilution. The concentration of these organisms in the open ocean may be no more than one-thousandth of that in laboratory cultures. An accurate evaluation of the flow of energy and matter through the ocean's biomass requires measurements of physiological activities as well as of the numbers and kinds of the individual organisms. Methods currently employed in the isolation of phytoplankton are inadequate: filtration or continuous-flow centrifugation often encourages disintegration of the more delicate species or their masking by organic debris. Lammers [1], using centrifugation in density gradients of iodinated benzoic acid, succeeded in isolating *Chlorella vulgaris* which maintained photosynthetic competency, but the gradient material had to be removed from the algae slowly and carefully.

Most of the materials used in gradient preparation preclude the isolation of physiologically active marine algae. Toxicity is the primary drawback of salts of heavy metals. Sucrose elevates the osmotic pressure sufficiently to cause plasmolysis with a resulting increase in cell density. Ficoll is not dense enough to prevent pelleting of the microorganisms. The viscosity of both polymers, moreover, makes their use in centrifugations of this kind impractical. In experiments previously reported from our laboratory [2], some fairly tolerant marine algae were sorted according to buoyant density in gradients of the silica sol Ludox AM. Taking into account that Ludox AM exhibits low viscosity and osmotic potentials, gelation occurs at saline concentrations which are half that of seawater.

The introduction by Pharmacia of 'Percoll', a silica sol co-valently bonded to polyvinylpyrrolidone, led to the evolution of the method we now describe [3]. Percoll exhibits low osmotic potentials and electrophoretic mobilities, has a density of 1.13 g/ml, which can be modified upward by dialysis or centrifugation, and is essentially non-toxic. Percoll differs from Ludox HS (from which it is prepared) and from Ludox AM by virtue of the masking of the surface changes common to silica particles (Fig. 1). Whiskers, terminating in PVP, sequester or suppress the negatively charged oxygens on the silica surface [4].

PROCEDURES

The osmotic basis of the high and low density solutions which are used to construct the gradients is 0.5 M sorbitol (Table 1). Both linear and step gradients have been employed. While this sorbitol-seawater mixture (SSW) is not osmotically or compositionally equivalent to natural seawater, it is compatible with a number of algae. Its addition to Percoll (PSSW) provided sufficient density to permit isopycnic banding of all of our test organisms.

Separations were carried out as follows: 85 ml of a culture with 15 ml of the gradient placed underneath was centrifuged at 2000 rev/min for 20 min. To avoid perturbing the gradients it was necessary to accelerate and decelerate slowly, between 0 and 300 rev/min. It can be seen from Figs. 2 and 3 that the bands of cells range from the bottom to the upper third of the gradient, depending on the species. Fig. 3 illustrates a separation of *Phaeodactylum* and *Nitzschia* species from a sewage pond (tube 2) in which the banding density of *Phaeodactylum* is identical to that obtained by sedimentation of the pure culture in tube 1. These and other diatoms, dinoflagellates, brown, and green algae, and a species of blue-green bacterium, were all selected from the Woods Hole culture collection for their fragility or size.

Table 1. Composition of gradient solutions. Amounts are calculated for 1 litre vols. of solution. *From ref. [3].*

	Component	Amount	Final concentration
Starting solution of low density, viz. 1.03 (SSW)	sorbitol	91.1 g	0.5 M
	Tris-HCl	2.65 g	0.05 M Tris, pH 8.5*
	Tris-base	4.03 g	
	$MgCl_2$	1.43 g	0.015 M
	seawater	100 ml	10% v/v
	distilled water	to 1000 ml	
Final solution of high density, viz. 1.15 (Percoll-SSW, purified; added in 2 portions)	sorbitol	91.1 g	0.5 M
	Tris-HCl	2.64 g	0.05 M Tris, pH 8.5[+]
	Tris-base	4.03 g	
	Percoll	500 ml	50% v/v
	$MgCl_2$	1.43 g	0.015 M
	seawater	100 ml	10% v/v
	Percoll	to 1000 ml	silica final concn. ∿18% w/v

* *Olisth* required buffering at pH 7.5 instead of 8.5

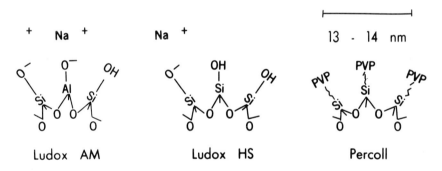

Fig. 1. Surface configuration of silica sols.

Upon recovery from the gradients, most of the algae were
motile, although slowed somewhat by the viscosity of the medium.
An exception was *Peridinium trochoidium* which was uniformly
rounded and inactive. Cell counts in a Sedgewick-Rafter chamber
showed the process of centrifugation itself to be deleterious to
motility in this dinoflagellate, while the gradient material
merely reduced motility by 30% (Table 2).

All of the algal species shown were diluted after centrifugation and inoculated into fresh media. After one week, live
organisms were recovered from all samples. Motility appeared
normal and growth ranged from slight to moderate. The recovery

Fig. 2 and 3 *(on right, below)*. Isopycnic sedimentation of marine phytoplankton into gradients of Percoll in SSW. Linear gradients 0 to 90% v/v Percoll in SSW were inserted under samples of algal cultures or natural waters in 100 ml centrifuge tubes, and centrifuged at 2,000 rev/min for 20 min.

Fig. 2. — *From left to right:* (1) Pond A, a culture predominantly of *Phaeodactylum tricornutum* growing in a sewage-enriched pond of seawater; (2) Pond B, a mixed culture of *Phaeodactylum* and *Nitzschia* species growing in a sewage-enriched pond of seawater; (3) culture of clone *Skeletonema;* (4) culture of clone of 7-15 of *Thalassiosira pseudonana*.

Fig. 3. — Cultures of laboratory clones, *from left to right:* Peri, 13-10 Pyr, Iso, SM-24 *(Oscillatoria)*, T-13-L *(Cyclotella)*, and 3-H. *From ref. [3].*

Table 2. Effect of centrifugation or Percoll in SSW on the motility of *Peridinium trochoideum*. Untreated culture was counted directly by microscopic examination, then after centrifugation in seawater alone as described. An equal volume of PSSW was added to fresh control culture to determine the effect of the gradient medium. *From ref. [3].*

Treatment	% Motility
Untreated culture	65.2
Centrifuged in seawater	0.0
Culture diluted to 50% v/v with PSSW	35.9
Centrifuged in gradient of PSSW	0.0

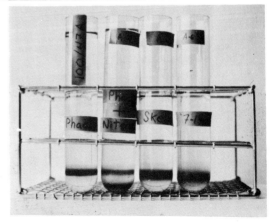

of photosynthesis after aspiration from the gradient was tested with laboratory cultures of *Olisthodiscus luteus* (Table 3). Here, two bands separated in the gradient, corresponding to a major recovery of intact cells and a very dilute band. Photosynthesis, measured with the oxygen electrode, was determined with both fractions and dilutions therefrom and was compared with untreated cultures. Net photosynthetic activity of the recovered cells remained undamaged. In contrast, when cells were pelleted by ordinary centrifugation, irreversible aggregation occurred which precluded any meaningful measurement.

Table 3. Recovery of photosynthesis by *Olisthodiscus luteus* after centrifugation in Percoll gradients. The chlorophyll concentration was normalized at 5.75 µg/ml. *From ref. [3].*

Fraction	PS Rate (nmol O_2/min/ml)
Control	5.3
Gradient top	6.9
Gradient mid	4.8

CONCLUSION

Centrifugation in gradients of Percoll permits the collection, concentration and resolution of marine species from laboratory cultures. Most of these organisms are recovered in a condition that would permit direct physiological measurements of photosynthesis, respiration, ion absorption and specific growth rates.

Acknowledgements

These studies were supported in part by grant PCM 77-04674 from the National Science Foundation (U.S.A.) and the Charles and Johanna Busch Memorial Fund.

References

1. Lammers, W.T. (1971) in *Water and Water Pollution Handbook* (Ciaccio, L.L., ed.), Vol. 2, Marcel Dekker, New York, pp. 593-638.
2. Price, C.A., Mendiola-Morgenthaler, L.R., Goldstein, M. & Guillard, R.R.L. (1974) *Biol. Bull. 147*, 136-145.
3. Price, C.A., Reardon, E.M. & Guillard, R.R.L. (1978) *J. Limn. & Ocean. 23*, 548-553.
4. Pertoft, H. & Laurent, T.C. (1977) in *Methods in Cell Separation,* Vol. 1 (Catsimpoolas, N., ed.), Plenum, New York, pp. 25-65.

#D-2

PROBLEMS IN THE SEPARATION OF PARASITIC PROTOZOA

SHEILA M. LANHAM

Department of Medical Protozoology,
London School of Hygiene and Tropical Medicine,
London, WC1, U.K.

*The protozoa are unicellular eukaryotes and the smallest animals
It is becoming increasingly important, in the search for methods
of control of the parasitic protozoa pathogenic to man and animals, to be able to isolate the organisms as free as possible
from the contaminating tissues of the vertebrate host and correspondingly from any vectors which may be involved in the life-
cycle. During the successive stages of their often complex life-
cycle, these protozoa exhibit morphologically different forms,
which may require different separation techniques. Thus, the
malaria parasite shows 12 different main stages of development,
7 in the vertebrate host and 5 in the mosquito.*

*Consideration is given to methods and problems involved in isola-
ting parasitic protozoa of medical and veterinary importance be-
longing to the genera Trypanosoma, Leishmania, Plasmodium, Eimeria
or Toxoplasma. Thus, for Trypanosoma cruzi in a vertebrate host
the flagellate bloodstream forms can suitably be isolated by a
DEAE-cellulose column, and non-flagellate intracellular forms
(muscle) by homogenization followed by DNase-collagenase-trypsin
and sucrose density-gradient centrifugation.*

Protozoa of the genera *Trypanosoma, Leishmania* and *Plasmodium*
undergo a cycle of development in the vertebrate host and a
further cycle of development in an insect vector prior to transmission to a new vertebrate host. Exceptions to this pattern are
some trypanosome species adapted to mechanical transmission.
Toxoplasma and *Eimeria* spp. have no insect phase and rely on
contaminative transmission of infective cysts and/or oocysts between
vertebrate hosts for their perpetuation. Protozoa also vary
notably in the complexity of their reproductive stages. *Trypanosoma* and *Leishmania* spp. reproduce only asexually during their
life-cycles with relatively few changes in morphology, whilst
Plasmodium, Eimeria and *Toxoplasma* spp. reproduce both sexually
and asexually, passing through many developmental stages which
exhibit gross differences in morphology, ultrastructure and
metabolism. The sites of development of the many protozoal
stages in the host and vector (if present) are either extracellular

or intracellular, and the task of separating out the different
parasite stages is a formidable one, hitherto not yet feasible
for every stage of some of the above protozoa.

The development of *in vitro* culture methods has greatly faci-
litated the isolation of certain stages of most of the above
protozoa [see #1, 2 & 4-7 in 1], although with those cultivated
intracellularly the problem of isolation from the host cell remains. Where practicable, however, organisms cultured *in vitro*
should be compared with those of the same stock isolated from an
infected vertebrate host or vector, to disclose any divergence
in intrinsic or extrinsic characters. For this reason and
because of the many parasite stages still refractory to *in vitro*
cultivation, procedures must still be sought for the isolation
of protozoa from the infected vertebrate host and vector.

Before laboratory hosts can be infected for the purpose of
harvesting parasites in quantity,   there are often problems to
resolve, e.g. low infectivity of some protozoal species and
restrictive host specificity. Some species adapt well to laboratory animals after serial sub-passage, although adaptation may
be restricted to certain strains of a species as shown by
Trypanosoma vivax in rodents [2]. Often the common laboratory
animals are refractory to infection, or only exhibit a mild
chronic disease with low parasite levels. Parasite numbers can
be increased by reducing the host's immune response, as by γ-
irradiation in rodents infected with *T. cruzi* [3], by cyclophos-
phamide as an immunosuppressant in *T. brucei gambiense* infections [4], and by use of splenectomized monkeys for some
Plasmodium species [5]. Exploring the field of unusual vertebrates to find more susceptible host species has proved fruit-
ful, e.g. the use of the multimammate rat, *Mastomys natalensis,*
for *T.b. gambiense* [4], the chinchilla *(Chinchilla laniger)* for
T. cruzi [3] and the owl monkey *(Aotus trivirgatus)* for human
pathogens *(Plasmodium falciparum, P. malariae* and *P. vivax)*
which are non-infective to rodents and to the common guenon and
macaque monkeys [5].

DESIGN CONSIDERATIONS FOR A SEPARATION METHOD

The main requirements in an isolation method for protozoa are
for the operator to be able to harvest sufficient numbers of
viable organisms with minimal host-cell contamination, over a
convenient time period and as simply as possible. Sometimes it
is difficult to attain or combine the criteria of purity, viability and quantity. This is especially so where a parasite stage
in vivo is naturally short-lived, e.g. the merozoite of malaria
[6] or when the numbers of parasites present in the host or
vector are characteristically low and/or intracellular. A
further problem is when the different parasite stages are mixed,

e.g. the asexual and sexual stages of malaria in asynchronous infections.

Sometimes the isolation method has to serve, with requisite sensitivity, for parasitological diagnosis and morphological identification of an organism in scanty infections when standard identification methods have failed [7].

Cryopreservation for the long-term storage of viable organisms is an important aid to separation studies and to protozoal research in general [8]. It provides a means of preserving freshly isolated material and stocks in general, thereby eliminating the need for maintenance by serial passage in animals or *in vitro* culture, with the concomitant risk of changes in the characteristics of the organisms [9].

SEPARATION METHODS

The isolation of individual species of particular veterinary and medical importance and those used as laboratory models will be discussed under their respective genera. Readers unfamiliar with the characteristics of the species concerned are recommended to consult references [10-12] for general information, [13] for detailed information on trypanosomes and similarly [14, 15] for malaria.

Trypanosoma

Trypanosomes infecting mammals are classified by their station (locus) of development in the insect vector into two sections, *viz.* the *Stercoraria* and the *Salivaria*. There are four main developmental stages in trypanosomes: three morphologically different flagellate forms termed trypomastigote, procyclic trypomastigote and epimastigote, and the rounded non-flagellated amastigote. Transitional forms link these various stages. In the *Salivaria* the trypomastigotes are the dividing forms found in the bloodstream, whilst the procyclic tryponastigotes and epimastigotes occur only in the insect vector. In the *Stercoraria*, with certain species, e.g. *Trypanosoma cruzi,* the dividing forms in the host are the amastigotes which reproduce intracellularly, then transform *via* epimastigotes to blood trypomastigotes. All four forms are present in the insect vector.

Section *Salivaria*

SEPARATION OF BLOOD FORMS

The major species in this section are (a) *T.b. rhodesiense* and *T.b. gambiense* infective to man and animals, and (b) *T. vivax, T. congolense, T. simiae, T.b. brucci, T.b. evansi* and *T.b. equiperdum* infective to animals only. All species of salivarian trypamosomes live and multiply as trypomastigotes in the bloodstream and lymph of the mammalian host, but some of the *T. brucei*

subspecies also exist and proliferate in the subcutaneous and
interstitial tissues. Separation methods have been applied to
isolate the organisms from the blood of a variety of hosts
ranging from the usual laboratory rodents to domestic stock such as
cattle, goats, sheep, equines, dogs, pigs and camels. Earlier
methods relied on differential centrifugation (differential
pelleting) followed by removal of contaminating blood cells by
agglutination with a phytohaemagglutinin or specific anti-serum
[16]. The disadvantages of these methods were that (a) preparations were frequently contaminated with blood cells and platelets [17], (b) they were time-consuming with a consequent dele-
trious effect on the organisms, and (c) certain species, e.g.
T. congolense, had a density very close to that of the host
erythrocytes and could not be separated by simple centrifugation.
A later procedure with a sucrose density gradient gave much
purer preparations but was restricted to small volumes of blood
and required a preliminary defibrination to facilitate the removal of platelets [18].

A more recent method was that of column separation of infected blood on the anion-exchanger DEAE-cellulose (DE52, Whatman),
the principle of the method being that the very negatively
charged blood cells were retained by the exchanger, but not the
less negatively charged trypanosomes [19]. The ionic strength
(I) of the pH 8.0 phosphate-saline-glucose buffer (PSG) used as
diluent and for equilibrating the DE52 was critical for each
host-blood species. Erythrocytes with a low negative surface
charge, e.g. pig and rabbit, required PSG of a much lower ionic
strength (I = 0.11) than, e.g., rat or horse erythrocytes with a
higher negative surface charge, where PSG of I = 0.22 was used.
It proved inadvisable to decrease the ionic strength of buffers
below that necessary for satisfactory retention of the blood
cells, as the trypanosomes themselves were liable to be retained
at the lower ionic strengths. The harvested trypanosomes showed
no contamination with blood cells or platelets, and up to 200 ml
of blood could be processed on a column of appropriate size.
All the salivarian trypanosome species were successfully separated, most of them from a wide range of infected hosts. The
method has gained wide acceptance and gives the most satisfactory
preparations with regard to purity, viability and yield of any
tested so far. Because of its high sensitivity, the method was
also used in the diagnosis of Gambian sleeping sickness [7, 20],
and has recently led to a device consisting of a small diagnostic
column for a 50 µl blood sample [21]. A further development was
the separation of *T.b. gambiense* from homogenized infected mouse
tissues [22].

SEPARATION OF CULTURE FORMS

The tsetse-transmitted species of *Trypanozoon* and *T. congolense*
can be grown in cell-free media and recovered by simple centrifugation [23]. The haematozoic forms transform and multiply mainly

as the so-called procyclic trypomastigotes which correspond to
the mid-gut forms in the tsetse fly, and in this respect offer a
useful substitute for the insect phase. They are non-infective
to vertebrates differing from the parent haematozoic forms in
metabolism and ultrastructure, and are strongly held by DEAE-
cellulose (DE52, Whatman) — a property that was utilized to
separate them from non-transformed organisms present during early
cultivation stages (*pers. comm.* from Dr. D.A. Evans of this
School).

Recently, infective forms of *T.b. brucei* (strain 427) were
grown *in vitro* for the first time over bovine fibroblast-like
cells in modified RPMI 1640 medium [24]. Similar results were
obtained using buffalo and Chinese hamster lung-cell lines [25].

Section *Stercoraria*

Stercoraria are typically non-pathogenic and of restricted host
specificity, with the exeption of *T. cruzi,* the causative organism of Chagas' disease in man, which is transmitted by reduviid
bugs.

T. cruzi

SEPARATION OF BLOOD AND TISSUE FORMS

In the vertebrate host the organisms invade tissues, especially
reticuloendothelial cells, heart muscle and skeletal muscle,
transforming into rounded non-flagellated amastigotes which
multiply intracellularly by binary fission forming pseudocysts.
After some days the amastigotes metamorphose to non-dividing
flagellated trypomastigotes which are dispersed into the bloodstream ready to repeat the cycle of tissue invasion.

The separation of the haematozoic and tissue forms of *T. cruzi*
has hitherto presented a problem; but a satisfactory and econo-
mical method for their recovery was recently devised by Gutteridge *et al.* [3] whereby both forms were recovered in parallel
from the same infected animals. A precise protocol was required
for the production of heavy blood and tissue infections, and
likewise for the recovery of the organisms, and the very explicit
original details should be consulted. Chinchillas, mice and γ-
irradiated rats were infected with the Sonya strain of *T. cruzi*
maintained in mice. Separations were carried out when high
parasitaemias correlated with high levels of releasable intracellular forms, and are summarized below.

Trypomastigotes were separated by a considerable modification
of the DEAE-cellulose column technique for salivarian trypanosomes [19]. An important step was the preliminary low-speed
centrifugation of infected blood diluted with a high ionic-
strength citrate — PSG anticoagulant solution. The trypanosomes in the supernatant were then separated on a shallow DEAE-

cellulose column using a PSG buffer, pH 7.5, I = 0.206.   Because
of the characteristic fragility of *T. cruzi* trypomastigotes, a
concentration of 10% (v/v) heat-inactivated serum from newborn
calves was maintained in the eluate to preserve the organisms
undergoing elution.   All procedures were carried out at room
temperature, and eluted organisms were centrifuged and washed at
⊁1000 g.   The harvested trypomastigotes showed no loss of infectivity to mice, and yields were $3 - 6 \times 10^8$/rat and 1×10^9/chin-
chilla.

For the recovery of amastigotes, hind muscles were cut up
into small pieces in a special breakage medium, washed and then
homogenized in an MSE homogenizer (all at 4°).   The gauze-filtered
suspension was treated at room temperature with DNase (5 min),
then with collagenase and trypsin (20 min), and filtered.   The
amastigotes were recovered from the filtrate by a precise differential centrifugation procedure.   Contamination by red cells
was 1-2%, and by white cells 2-10%, this being reduced for each
to 0.1% by sucrose density-gradient centrifugation.   Average recoveries of organisms were 1×10^8/rat and 1×10^9/chinchilla; of
these, 1-3% were epimastigotes plus trypomastigotes, and the re-
mainder amastigotes.

Trypomastigotes have also been isolated from the peritoneal
fluid of infected mice using a column of Cellex-SE cellulose
(Bio-Rad Labs., Richmond, Calif.) [26].

SEPARATION OF IN VITRO CULTURE FORMS

Epimastigote, trypomastigote and amastigote forms can be grown
in cell-free or tissue-culture media, the proportions of the
different forms obtained depending on the type of medium and the
incubation temperature [23, 27].   A recently devised highly
enriched liquid medium supported amastigote growth at 37° (hitherto only possible at 35°) [28].   DEAE-cellulose columns were used
to separate mixtures of epimastigotes, trypomastigotes and amastigotes [29, 30].   Amastigotes and trypomastigotes could be eluted,
whilst the epimastigotes were retained on the anion-exchanger.
Others found complete retention of all forms on DEAE-cellulose,
whereas with CM-cellulose amastigotes were retained and trypomastigotes eluted [31].   These conflicting results were perhaps
due to the use of different *T. cruzi* strains and/or the use of
different brands of DEAE-cellulose, which can differ in ion-
exchange characteristics.

T. lewisi

SEPARATION FROM BLOOD

This non-pathogenic species frequently used as a laboratory model
was separated from infected rat blood by a modification of the
DEAE-cellulose column technique [32].   The organism is easily
cultured in cell-free media [23].

Leishmania

The main species infecting man are *Leishmania donovani*, the cause of visceral leishmaniases and *L. tropica*, *L. mexicana* and *L. braziliensis* which cause cutaneous leishmaniases. The parasite is transmitted by sandflies (*Phlebotomus* spp.) as a long slender flagellated promastigote. On entry into the vertèbrate host the parasite immediately becomes established in the cells of the reticuloendothelial system; but unlike *T. cruzi*, which has a blood form, the leishmania exists only in the amastigote form, multiplying by binary fission in macrophage-type cells.

SEPARATION OF VARIOUS FORMS

Human species are infective to laboratory rodents, and amastigotes of the cutaneous forms can be isolated from skin lesions. In a recent rapid method (40 min), amastigotes of *L. mexicana amazonensis* were recovered by filtering the supernatant from skin-lesion homogenate through a dry cellulose (Whatman CF11) column. Viable amastigotes were eluted with 0.5-1.0% red cells, and if necessary were further purified on a linear sucrose gradient [33]. In a similar method with the same species, amastigotes were isolated by pouring the homogenate supernatant through a column of glass beads (0.1 mm diam.) [34]. In more complicated procedures, skin lesions from animals infected with *L. mexicana* or *L. braziliensis* (or *L. enriettii*, which does not infect humans) were trypsinized and then incubated for 18-24 h in a tissue-culture medium, after which relatively pure amastigotes were recovered from the liquid overlay [35].

Culture forms equivalent to the promastigote insect phase grow easily in simple culture media [23], whilst the amastigote forms grow intracellularly in cultured cells [27].

Plasmodium

Medically the most important species are *Plasmodium vivax*, *P. malariae*, *P. ovale* and *P. falciparum*, this last species being the most pathogenic to man. Other species used frequently in research because of the narrow host-specificity of the human malarias are *P. knowlesi*, *P. cynomolgi* and *P. coatneyi*, which are infective to monkeys; *P. berghei*, *P. yoelii*, *P. vinckei* and *P. chabaudi*, infective to murine rodents; and the avian malarias *P. gallinaceum* and *P. lophurae*.

The first sites of asexual reproduction of the parasite in the mammalian host are the *primary exo-erythrocytic schizonts* formed on invasion of the liver parenchymal cells by sporozoites introduced into the host *via* the bite from an infected mosquito. Each liver schizont yields 10,000-30,000 merozoites which pass into the blood stream, invading the erythrocytes to give the *intra-erythrocytic forms*.

Blood stages. — With many species these present a heterogeneous
mixture of extra-cellular merozoites and parasitized erythrocytes
and possibly reticulocytes.   A small proportion of the
intra-erythrocytic forms consists of the sexual forms of the
parasite, i.e. the micro- and macro-gametocytes, whilst the re-
mainder is made up of the asexual forms at various stages of
development.   These intracellular asexual forms range from the
ring forms of the recently invaded erythrocyte, to the developing
trophozoite forms and finally the schizont and segmenter stages,
where up to 6 mitoic divisions may have occurred within the
erythrocyte to form the merozoites for release into the blood and
subsequent re-invasion of the erythrocytes.   The degree of
heterogeneity found amongst the asexual forms in the peripheral
blood depends on two main factors, (a) whether or not the infection is synchronous, and (b) whether the development of the
schizont stage is restricted to the deep vascular sites, through
endothelial adhesion as in *P. falciparum*.   These two factors and
subsidiary ones affecting the parasitaemias, e.g. the predilec-
tion of some species for reticulocytes, are in turn dependent on
a combination of the characteristics of the strain and species of
the malaria and the species of the host.

Mosquito stages. — The male and female gametocytes are ingested
from the blood into the stomach of the female mosquito, where
the sexual cycle is completed with the formation of the ookinete
which burrows through the stomach wall to form an oocyst on its
outer surface.   Here thousands of sporozoites are produced which
migrate to the salivary glands ready for transmission to the
vertebrate host.

Currently, most stages of the parasite are being investigated as potential vaccine material [36], in relation to the
demand for isolated organisms with minimum contamination by host
material.   As Kreier [37] recently gave an extensive critical
review of isolation methods the following discussion will
emphasize methods subsequently described.

SEPARATION OF BLOOD FORMS

Removal of leucocytes and platelets. — This prerequisite to the
isolation of malarial blood forms [38-40] usually entails (a)
the use of cellulose columns and glass beads, and/or (b) sedimentation procedures.   Filtration through cellulose powder columns removed nearly all the leucocytes from *P. berghei*-infected mouse
blood and was superior to sedimentation procedures [41].   In one
method platelets were removed prior to cellulose treatment by
defibrination with glass beads (5 mm diam.) [42].   In another,
the platelets were aggregated by the addition of ADP to washed
P. berghei-infected blood prior to filtration through a column of
glass beads overlaid with cellulose powder (Whatman, CF11) [43].
Similar mixed columns removed >99% of lymphocytes, granulocytes,

platelets, and haemopoietic stem cells from pooled blood and
spleen suspensions from P. berghei-infected mice [44]. Some
workers have preferred filtration through a dry cellulose column,
whilst others have advocated pre-wetting [37]. Recently columns
of sulphoethyl-cellulose (SE-23, Serva, Heidelberg, W. Germany)
mixed with Sephadex G25 (Pharmacia) proved successful for the
removal of platelets and leucocytes from P. chabaudi-infected
mouse blood, prior to zonal separation of parasitized from non-
parasitized erythrocytes [45], and with P. berghei-infected
blood prior to schizont purification using Ficoll-Paque (Pharmacia) mixtures [40].

Schizont and merozoite separation. — Merozoites are released
into the blood on maturation of liver and blood schizonts. The
number of liver schizonts formed in vivo and their resulting
merozoites are too few to allow of isolation. In vitro systems
for the production of exo-erythrocytic (EE) schizonts and merozoites have so far been successful only with avian malarias, for
reasons discussed by Beaudoin [46]. Methods discussed here will
be for the erythrocytic schizonts and merozoites. With some
malaria species asynchronism and/or the preference of blood
schizonts for the deep vasculature mitigate against good recovery
of merozoites from freshly drawn blood. This problem generated
separation methods based on a preliminary in vitro maintenance
of parasitized erythrocytes through to the mature schizont stage,
with consequent 'natural release' of merozoites and subsequent
recoveries by methods best suited to the particular parasite
species - host species combination. Other methods have entailed
artificial release, e.g. lysis or ultrasonication. Both approaches were reviewed by Kreier [37], and methods subsequently
described are outlined below. Centrifugation methods for schizont concentration were briefly reviewed in a report which gave
a method utilizing spleen homogenates as well as blood from P.
berghei-infected mice [44].

In a comparative study of five methods of artificial release,
all resulting merozoite preparations were non-viable [47]. A
further study indicated that late schizonts and merozoites were
labile at low temperature and sensitive to centrifugation; loss
of ability to invade red cells was complete within 10-15 min
after release, in accord with the earlier findings (cited in
[6]). Siddiqui et al. [48] obtained an effective experimental
vaccine of crude merozoites and segmenters from P. falciparum-
infected Aotus monkey blood by in vitro maintenance (35-40 h)
followed by saponin lysis and density-gradient centrifugation.
Another effective vaccine was prepared from naturally released
merozoites of P. falciparum isolated from cultures of schizont-
infected human red cells maintained on perfused columns of
cellulose (Whatman CF11) [49]. Recently, purified viable merozoites were recovered from P. chabaudi-infected mouse blood for
the first time by natural release [50]. Inversion of the

nycthemeral cycle of the mice converted the nocturnal schizogony peak into a more convenient diurnal one. Blood at peak
schizogony was perfused through a column of Con-A-Sepharose
(Pharmacia) at 37°. Merozoites were eventually eluted as a peak
containing <0.1% red cells, ≮6% immature parasites and no leucocytes or platelets. Another method (see H.-G. Heidrich & K.
Hannig, this vol.) entailed free-flow electrophoresis to separate parasitized from non-parasitized erythrocytes of P. vinckei-
infected blood and to purify free parasites after mechanical release from parasitized erythrocytes agglutinated by concanavalin-
A [51].

SPOROZOITE SEPARATION

As with most malarial stages there was an urgent requirement for
the isolation of pure sporozoites from mosquitoes for use with
in vitro techniques [46] and for experimental vaccines. The
usual preliminary isolation steps were trituration in medium or
buffer of either the whole mosquito or particular parts containing sporozoites, e.g. mid-gut with oocysts, thorax (containing salivary glands and haemocele), or salivary glands alone.
The suspension of sporozoites, mosquito debris and microbial
contaminants was cleared of large particles by slow centrifugation, and the sporozoites further purified by density-gradient
procedures. A recent example was the use of a biphasic gradient
of bovine serum albumin-Diatrozoate mixtures in 'medium 199'
[52]. As a completely new approach, like that for trypanosomes
[19], a column separation has been performed with DEAE-cellulose
[53, 54]. Sporozoites isolated by this speedy technique retained their motility, infectivity and immunogenic properties,
whilst fungal contamination was eliminated and bacterial contamination considerably reduced. P. knowlesi sporozoites showed
a different column elution profile to those of P. berghei and
P. cynomolgi [53].

Eimeria

Species of this genus and of Toxoplasma form part of the
'coccidia' group, the former species causing coccidiosis in
animals and the latter toxoplasmosis in man. Most Eimeria
species are intestinal parasites and vacate the host via the
faeces in the form of resistant oocysts. The very resistant
oocyst wall permits the use of various chemicals such as dilute
solutions of potassium dichromate, H_2SO_4 and hypochlorite and
saturated solutions of NaCl for recovering oocysts from faeces.

SEPARATION OF VARIOUS FORMS

Ryley et al. [55] established a simple method for the recovery
of purified viable sporulated oocysts from E. tenella-infected
chicken faeces. Sporocysts are usually released from oocysts
by mechanical breakage of the oocyst wall, e.g. by a Teflon-
coated tissue grinder or by shaking with glass beads. In vitro
excystation of sporozoites from sporocysts is effected by

incubating with bile salts and trypsin [27]. Chymotrypsin present as an impurity in commercial trypsin is now thought to be the actual digesting agent [56]. Centrifugal elutriation using a JE-6 elutriator rotor was successfully applied to the isolation of sporozoites, sporocysts and oocysts [57]. Second-generation merozoites of *E. tenella* were purified from the caeca of infected chicks and from the chorioallantoic membranes of embryonated eggs by hyaluronidase treatment followed by centrifugation in buffered 7.5% Ficoll-10% Hypaque solutions [58]. *In vitro* cultivation of the complete developmental cycle has so far only been possible with *E. tenella* [27].

Toxoplasma

Toxoplasma gondii is the major species and has an unusually wide range of host animals. The coccidian-type intestinal stages of schizogony and gametogony are restricted to the cat family which alone transmits infective oocysts *via* the faeces. The asexual forms reproduce intracellularly by endodyogeny and are of two types, the rapidly reproducing tachyzoites forming the pseudocysts of acute infections and the slower growing bradyzoites forming the true cysts of chronic infections. Experimentally, tachyzoites are conveniently recovered from the peritoneal exudates of acutely infected rodents, where they are found both free and in the macrophages.

SEPARATION OF TACHYZOITES

Fulton & Spooner released the intracellular forms by gently shaking the centrifugal deposit from the exudate with glass beads, and then purified the tachyzoites by filtration through a sintered glass filter of critical pore size [59]. Recoveries from a variety of rodents were compared, the highest yields being obtained from cotton rats. More recently, tachyzoites were recovered from infected mouse peritoneal exudate by density-gradient centrifugation in a zonal rotor (MSE type A) using gradients of either Ficoll or Dextran 40 [60].

CONCLUDING COMMENTS

Most of the separation approaches reviewed above have involved one or both of the following procedures: (a) differential pelleting and density centrifugation techniques similar to those used in whole-cell and subcellular fractionation; (b) selective retention (possibly adsorptive) of host-cell contaminants by column materials such as cellulose powders, ion-exchange celluloses, glass beads or Con-A-Sepharose, or by filtration through a graded glass filter. Group (b) procedures are often speedier and milder and have proved highly useful for the recovery of extracellular motile stages in a viable condition: e.g. the flagellated forms of trypamosomes, especially the fragile *T. cruzi* trypomastigotes;

malarial merozoites that have a limited life-span *in vitro* and
in vivo, and malarial sporozoites. These procedures should be
applicable to the separation of similar stages of other protozoa.

References

1. Taylor, A.E.R. & Baker, J.R., eds. (1978) *Methods of Cultiva-
   ting Parasites* in vitro, Academic Press, London. — *See
   particularly* Chapters 1, 2 & 4-7.
2. Leeflang, P., Buys, J. & Blotkamp, C. (1976) *Int. J. Parasit.
   6*, 413-417.
3. Gutteridge, W.E., Cover, B. & Gaborak, M. (1978) *Parasitology
   76*, 159-176.
4. Mehlitz, D. (1978) *Tropenmed. Parasit. 29*, 101-107.
5. Garnham, P.C.C. (1973) *Adv. Parasit. 11*, 603-630.
6. McAlister, R.O. (1977) *J. Parasit. 63,* 455-463.
7. Van Meirvenne, M., Janssens, P.G., Magnus, E. & Moors, A.
   (1973) *Ann. Soc. Belge Méd. Trop. 53*, 109-112.
8. Lumsden, W.H.R. (1972) *Int. J. Parasit. 2*, 327-332.
9. Baker, J.R. et al. (1978) *Bull. World Health Org. 56*, 467-480. — *A report.*
10. Baker, J.R. (1973) *Parasitic Protozoa, 2nd edn.,* Hutchinson,
    London.
11. Levine, N.D. (1973) *Protozoan Parasites of Domestic Animals
    and Man, 2nd edn.,* Burgess, Minneapolis.
12. Smyth, J.D. (1976) *Introduction to Animal Parasitology, 2nd
    edn.,* Hodder & Stoughton, London.
13. Hoare, C.A. (1972) *The Trypanosomes of Mammals, 1st edn.,*
    Blackwell, Oxford.
14. Garnham, P.C.C. (1966) *Malaria Parasites and Other Haemo-
    sporidia,* Blackwell, Oxford.
15. Killick-Kendrick, R. & Peters, W., eds. (1978) *Rodent Malaria,*
    Academic Press, London.
16. Simmons, V., Knight, R.H. & Humphryes, K.C. (1964) in *Publ.
    No. 97, 10th Int. Scienti. Cttee. for Trypanosomiasis Res.,
    Kampala,* p. 81.
17. Dixon, H. (1966) *Nature (Lond.) 210,* 428.
18. Williamson, J. & Cover, B. (1966) *Trans. R. Soc. Trop. Med.
    Hyg. 60,* 425.
19. Lanham, S.M. & Godfrey, D.G. (1970) *Exp. Parasit. 28*, 521-
    534.
20. Godfrey, D.G. & Lanham, S.M. (1971) *Bull. World Health Org.
    45,* 13-19.
21. Lumsden, W.H.R., Kimber, C.D. & Strange, M. (1977) *Trans. R.
    Soc. Trop. Med. Hyg. 71,* 421-424.
22. Lumsden, W.H.R., Kimber, C.D. & Doig, S. (1978) *Trans. R.
    Soc. Trop. Med. Hyg. 72,* 214.
23. Evans, D.A. (1978), *as for* [1], pp. 55-88.
24. Hirumi, H., Doyle, J.J. & Hirumi, K. (1977) *Science 196,*
    992-994.
25. Hill, G.C., Shimer, S., Caughey, B. & Sauer, S. (1978) *Acta
    Trop. 35,* 201-207.
26. Howells, R.E., Chiari, C.A. (1973) *Trans. R. Soc. Trop. Med.
    Hyg. 67,* 723-724.
27. Ryley, J.F. & Wilson, R.G. (1978), *as for* [1], pp. 111-128.
28. Pan, S.C. (1978) *Exp. Parasit. 45,* 215-224.

29. Al-Abbassy, S.N., Seed, T.M. & Kreier, J.P. (1972) *J. Parasit. 58,* 631-632.
30. Kreier, J.P., Al-Abbassy, S.N. & Seed, T.M. (1977) *Rev. Inst. Med. Trop. São Paulo 19,* 10-20.
31. Kanbara, H., Fukuma, T. & Nakabayashi, T. (1977) *Biken J. 20,* 147-149.
32. Lanham, S.M. (1971) *Trans. R. Soc. Trop. Med. Hyg. 65,* 248-249.
33. Brazil, R.P. (1978) *Ann. Trop. Med. Parasit. 72,* 579-580.
34. Shaw, J.J. & Lainson, R. (1977) *J.Parasit. 63,* 384-385.
35. Childs, G.E., McRoberts, M.J. & Foster, K.A. (1976) *J. Parasit. 62,* 676-679.
36. Cohen, S. (1979) *Proc. R. Soc. Lond. B. 203,* 323-345.
37. Kreier, J.P. (1977) *Bull. World Health Org. 55,* 317-331.
38. Homewood, C.A. (1978) in *Rodent Malaria* (Killick-Kendrick, R. & Peters, W., eds.), Academic Press, London, pp. 169-211.
39. Williamson, J. & Cover, B. (1975) *Trans. R. Soc. Trop. Med. Hyg. 69,* 78-87.
40. Howard, R.J., Smith, R.M. & Mitchell, G.F. (1978) *Ann. Trop. Med. Parasit. 72,* 573-575.
41. Homewood, C.A. & Neame, K.D. (1976) *Ann. Trop. Med. Parasit. 70,* 249-251.
42. Eling, W. (1977) *Bull. World Health Org. 55,* 105-114.
43. McAlister, R.O. & Gordon, D.M. (1976) *J. Parasit. 62,* 664-669.
44. Sterling, C.R. (1978) *J. Parasit. 64,* 747-749.
45. Königk, E. & Mirtsch, S. (1977) *Tropenmed. Parasit. 28,* 17-22.
46. Beaudoin, R.L. (1977) *Bull. World Health Org. 55,* 373-376.
47. McAlister, R.O. & Gordon, D.M. (1977) *J. Parasit. 63,* 448-454.
48. Siddiqui, W.A., Kramer, K. & Richard-Crum, S.M. (1978) *J. Parasit. 64,* 168-169.
49. Mitchell, G.H., Richards, W.H.G., Butcher, G.A. & Cohen, S. (1977) *Lancet i,* 1335-1338.
50. David, P.H., Hommel, M., Benichou, J.-C., Eisen, H.A. & Da Silva, L.H.P. (1978) *Proc. Natl. Acad. Sci. U.S.A. 75,* 5081-5084.
51. Heidrich, H.-G., Rüssmann, L., Bayer, B. & Jung, A. (1979) *Z. Parasitenkd. 58,* 151-159.
52. Beaudoin, R.L., Strome, C.P.A., Mitchell, F. & Tubergen, T.A. (1977) *Exp. Parasit. 42,* 1-5.
53. Moser, G., Brohn, F.H., Danforth, H.D. & Nussenzweig, R.S. (1978) *J. Protozool. 25,* 119-124.
54. Mack, S.R., Vanderberg, J.P. & Nawrot, R. (1978) *J. Parasit. 64,* 166-168.
55. Ryley, J.F., Meade, R., Hazelhurst, J. & Robinson, T.E. (1976) *Parasitology 73,* 311-326.
56. Wang, C.C. & Stotish, R.L. (1975) *J. Parasit. 61,* 923-927.

57. Stotish, R.L., Simashkevich, P.M. & Wang, C.C. (1977) *J. Parasit. 63*, 1124-1126.
58. Stotish, R.L. & Wang, C.C. (1975) *J. Parasit. 61*, 700-703.
59. Fulton, J.D. & Spooner, D.F. (1957) *Trans. R. Soc. Trop. Med. Hyg. 51*, 123-124.
60. Masihi, K.N., Mach, O., Valkoun, A. & Jira,J. (1976) *Exp. Parasit. 39*, 84-87.

ISOLATION OF RED CELLS, LEUCOCYTES AND PLATELETS FROM BLOOD, with some comments about sub-population heterogeneity

N. CRAWFORD
Department of Biochemistry,
Institute of Basic Medical Sciences,
Royal College of Surgeons of England,
London, WC2, U.K.

Isolation procedures should be simple to perform, with separation media and conditions which are reproducible, and result in good purity of cell type with quantitative recovery. Cell integrity and viability should be maintained. Important considerations include the following. — 1) Choice of anticogulant. 2) Effects of exposure to foreign surfaces. 3) Temperature, pH and osmotic pressure effects. 4) Nutrients for maintaining metabolic health. 5) Other additives to inhibit undesirable changes. 6) Selective destruction of unwanted cell species.

For red cells *consideration has to be given to shape maintenance and the balance between intrinsic and extrinsic factors. For* leucocytes *red-cell sedimentation and lysis may have to be enhanced. Besides whole blood, peritoneal exudates can serve as a source. Approaches for* platelets: *differential sedimentation, gel filtration or selective adhesion to surfaces; consideration has to be given to maintenance of shape, energy charge, inhibition of specific release reaction, and to sub-populations.*

A major difficulty in research activities involving the formed elements of blood is to select the best procedure for isolating the particular population of cells of interest in a good yield, as a reasonably homogeneous cell pool and in a metabolically and functionally healthy condition. Just a few of the many procedures which are available for isolating specific blood cell species are now surveyed, and some of the principles upon which these procedures are based are discussed. No attempt has been made to document all the procedures currently available, and the examples which have been chosen generally reflect our own experience and perhaps some prejudice.

With almost all isolation methods some compromise is unavoidable, and the selection of a particular technical procedure becomes a personal one in which the predominant considerations are the experimental needs in terms of purity and total cell yield and the kind of information one is seeking, be it functional, analytical, enzymatic or immunological. A further complication, not widely recognized, is that even after the isolation of an apparently homogeneous cell population there may be considerable sub-population heterogeneity with respect to various parameters such as shape, volume, buoyant density, surface charge and metabolic and functional competence. At best such a population of cells will always represent a range of characteristics related to different levels of maturation and/or senescence as determined by their circulating *in vivo* life spans.

The major classes of the formed elements of blood are sufficiently different in density, size and volume (Table 1) that theoretically, by choosing the appropriate sedimentation conditions in a centrifugal field it ought to be possible to achieve reasonably good differential separations. However, that such simple fractionations rarely give satisfactory results is almost certainly due to each cell class not being represented as a uniform population. This matter of sub-population heterogeneity will be referred to again later.

Table 1. Density and volume of human blood cells.
Density data are from ref. [1], and volume data from [2] or, in the case of platelets, from R. Wallis *(pers. comm.)*.

Cell	Density, g/ml at $0°$	Volume, μm^3
Platelets	1.038-1.058	2-6
Erythrocytes	1.100	80-90
Lymphocytes	1.070	230
Neutrophils and other white cells [monocytes, basophils, etc.]	1.062-1.082	450-470

CHOICE OF ANTICOAGULANT

Of the three main methods of rendering blood non-coagulable, *viz.* defibrination, passage through ion-exchange columns and the addition of anticoagulants, the first two procedures can be largely discounted since considerable selective losses occur through trapping out of cells in the fibrin clot or in the exchange columns. The chemical anticoagulants most frequently used have been heparin, citrate, oxalate and EDTA. Heparin has been widely used as an anticoagulant in blood-cell separation

studies and has been frequently quoted as being the most inno-
cuous with respect to leucocyte integrity and function. It is
believed to act by inhibiting thromboplastin, but it also has
some anti-thrombin activity. Amounts of the order 10-20 u/ml
or 0.1-0.2 mg are generally regarded as sufficient to inhibit
coagulation. At these levels there is little evidence that it
is detrimental to red-cell function or metabolism; but platelet
aggregates are commonly seen in blood collected directly into
heparin, and hence it is generally not recommended for platelet
studies. There have been some reports of the effect of heparin
on certain enzymes and particularly RNase and DNase, but its
exact mode of action is not understood. Perhaps one reason for
the use of heparin is that unlike other anticoagulants it has no
effect upon blood Ca^{2+}. In some research activities this may be
advantageous since the relationships between the intra- and
extra-cellular pools of this and other divalent cations are less
disturbed, and there are many divalent cation-dependent metabolic
processes within blood cells.

In studies with platelets and particularly in the investigation of their propensity to aggregate in the presence of various
agents (ADP, collagen and adrenaline), the effects of heparin as
anticoagulant are not always reproducible and depend also on
whether it is the only anticoagulant or is present with citrate
[3, 4]. For example, aggregation of platelets induced by adrenaline and collagen is less well expressed in heparinized plasma
than in citrate alone [5, 6], whereas at lower levels of heparin,
0.5-2.0 u/ml compared with 5 u/ml or more, these properties are
unaffected [7, 8]. In some of the earlier preparations of heparin, contamination with other mucopolysaccharides was considerable, and this fact may be responsible for some of the literature disparity.

With regard to oxalate and citrate, both of which act by com-
plexing out calcium ions, the latter as a constituent of the
acid/citrate/dextrose mixture is used almost universally for the
collection of blood trans-
fusion blood. However,
careful pH control of the
anticoagulant blood mixture
is vital with citrate, and
Table 2 shows by how much
the buoyant density of red
cells can vary under different pH conditions. If
this pH-dependent density
effect is also true for
leucocytes and platelets,
then it is perhaps not sur-
prising that careful control of pH is important for reproducible
separations.

Table 2. Effect of pH on buoyant density of human red blood cells at 4°. Data from ref. [11].

pH	Density, g/ml
5.3	1.069
6.2	1.085
7.0	1.100

Opinions differ widely on the effect of acid/citrate/dextrose solutions on leucocytes, and although some workers do not recommend them for phagocytosis studies or any procedure in which motility testing is involved [9, 10], many biochemical and metabolic studies have been carried out with apparent success with transfusion blood cells.

K_2-EDTA, K_3-EDTA and Na_2-EDTA, all of which chelate Ca^{2+}, are widely used as anticoagulants and usually at concentrations of about 1 mg/ml blood. At this level there is little alteration of red cell size, no significant aggregation of platelets and minimal disruption of leucocytes.

TEMPERATURE, OSMOTIC PRESSURE AND VARIOUS ADDITIVES

Here there are very real differences of opinion in the literature, and even contributions by the same author are often not consistent. One difficulty is that the ideal conditions for the initial cell separation may not always be the most favourable ones for the subsequent handling and for the storage or preservation of the cells for viability or metabolic investigations. With regard to temperature, usually values between 6-12° are chosen, often on an empirical basis. In our experience, however, what seems to be most important is to avoid the rapid changes which can occur during centrifugation; a temperature-control device on the centrifuge is essential for good reproducible separations. It is perhaps worth pointing out that leucocytes and platelets contain microtubule structures concerned with shape maintenance and intracellular motile activities. These polymeric structures, which are calcium-labile, also tend to disassemble at low temperatures, and even though reassembly generally occurs on warming, they may not reorganize in the same sort of way due to compartment changes in the sub-unit protein pools and other cofactors essential for polymerization.

Osmotic pressure changes induced in blood by hypertonic or hypotonic conditions can have most profound effects upon integrity and function of the cellular elements. Most workers attempt to maintain conditions around 290-310 m osmol/1, and there is a whole range of balanced salt solutions mostly based upon modifications of Tyrode solution which have been used for re-suspending cells once they have been removed from their plasmatic environment. Information on these and on their particular uses is available elsewhere [appendices in 12, 13]. Double-distilled water should be used rather than ion-exchange resins. Most of the recommended solutions are autoclavable.

Many techniques involve the addition of other agents for specific purposes, the commonest being glucose to support metabolism and maintain the cell's adenylate energy charge. This is perhaps more important for the red cell than other blood cells, since its energetic requirements are fulfilled by glycolysis only and it contains no reserves of glycogen to draw on in adversity. Other additives such as proteolytic inhibitors, apyrase, adenosine, phosphodiesterase inhibitors, prostaglandins and cytochalasins, all generally added to maintain surface membrane integrity, have their enthusiasts; such cocktails are generally highly tailored to particular research needs. These variations are too numerous to refer to here, and the reader should consult the appropriate research papers describing interests analogous to his or her own.

ERYTHROCYTE ISOLATION

I will deal only briefly with the isolation of these cells, since apart from authors who have specific interests at the whole-cell level, e.g. in deformability, shape-change characteristics or membrane cation gradients and other transport fluxes, most of the red-cell literature concerns the use of either soluble lysates or the post-haemolytic residues known as red-cell ghosts, as reviewed [14]. There are well established procedures for these isolations preparatory to ghost preparation, and these are generally designed to avoid any shape changes or formation of surface irregularities (e.g. echinocyte transformation) that indicate a shift in the delicate balance between internal and external forces which might affect membrane integrity or cause a redistribution of membrane constituents. The isolated plasme membrane of the erythrocyte is perhaps the most studied surface membrane of all cells, and almost without exception blood collected into buffered acid/citrate/dextrose is the isolation procedure of choice, the plasma and buffy coat being removed after differential centrifugation and the erythrocytes then washed in either isotonic phosphate buffer or isotonic saline [15, 16].

Although it is well established that there is an age spectrum for red cells in the circulation, only rarely has this been taken into account in studies of membrane constituents, membrane properties, enzyme levels or general metabolic health of the red-cell population. It seems reasonable to assume that a cell without glycogen reserves and without ribosomes or mitochondria will show an age-related decline in metabolic activities, and that if not separated into defined age groups the measurement of any parameter will be an expression of the mixed population of cells. Such age-related fractionations have been carried out, usually either by graded lysis [17] or by density difference in albumin gradients [18, 19] or mixtures of albumin and phthalate esters [20]. There seems general agreement that red cells become smaller with increasing age, and this is supported by re-infusion/life-span data [21]. Moreover, even taking into account the possibility of a small centrifugal-field effect on cells separated in this way there does seem to be significant evidence of large chemical and enzymatic differences between red cells differing in density (and presumably age).

LEUCOCYTE ISOLATION

This topic can be considered under two general headings, the first concerned with their separation from whole blood samples and the second with the stimulated production of leucotyes in experimental animals and their isolation.

Whole-blood leucocyte isolation

One procedure which has dominated the early work in this field is the use of lytic agents to selectively destroy red cells and facilitate the isolation of other cell classes. Many such agents have been used *(see panel)*; but simple though this approach may be, concomitant damage to the leuco-cytes (and probably platelets too) almost certainly occurs, and morpho-logical and metabolic changes can generally be detected. If one con-siders, too, that the liberated haemoglobin can itself have a damaging effect upon other cells, these selective lytic procedures are perhaps best avoided.

AGENTS USED FOR THE LYSIS OF RED CELLS PRE-PARATORY TO LEUCOCYTE ISOLATION
_____

Acetic acid/tartaric
  acid
Lysolecithin
Gramicidin
Alcohol/water
Hypotonic saline
Anti-A or Anti-B haemo-
  lytic sera
Saponin
Streptolysin
Ammonium chloride
_____

It seems that the only situation in which a carefully controlled lysis can be used with any confi-dence is with sheep blood, since sheep red cells have an unusu-ally high osmotic fragility. By choosing the right conditions to destroy the sheep red cells it is possible to achieve a leucocyte survival rate, albeit somewhat selective, of up to 80%. Thus, in the Dain & Hall [22] procedure, 20 ml of blood is added to 188 ml of 40% buffered Ringer's solution. The mixture is left for exactly 5 sec and then iso-osmolarity restored by the rapid addition of salt solution. With this procedure mono-nuclear cells are preserved reasonably well and lymphocytes res-pond well to plant haemagglutinin. Polymorphonuclear leucocytes are, however, almost entirely destroyed.

Another approach which is possibly less damaging is that of sedimentation enhancement. If the blood already has a high ini-tial sedimentation rate, as is often the case in disease states such as leukaemia, then by a simple combination of 1 g sedimen-tation and some centrifugation steps it is possible to harvest considerable quantities of leucocytes in quite good morphologi-cal condition. With sedimentation procedures a fact not gene-rally recognized is that the sedimentation rate falls with in-creasing cell concentration since the channels for upward fluid

currents are narrower. High plasma/
cell ratios are necessary for efficient separations by sedimentation
with enhancing agents.

For acceleration of red-cell
sedimentation with normal blood
samples there are a variety of
agents, a number of which are
listed *(see panel)*. Fibrinogen
has been used by many workers [23-
29] and results in red-cell Rouleaux
formation and an enhanced sedimentation rate.

Some additional procedure is
generally necessary because it is
not always possible to remove completely the red cells from the
leucocyte preparation. The normal
ratio of red cells to leucocytes of
∿1000 : 1 can be reduced to 1 : 1 by

AGENTS WHICH INCREASE
SPEED OF RED-CELL SEDI-
MENTATION

----

1. Fibrinogen
2. Gamma globulin
3. Dextran (mol. wt.
   $140-230 \times 10^3$)
4. Plant haemagglutinins
   (e.g. PHA)
5. Polyvinyl pyrrolidone
   (PVP)
6. Gelatin
7. Polybrene (hexadime-
   thrine bromide)
8. Methylcellulose
9. Isopaque (mixed with
   Ficoll)
10. Ficoll (mixed with
    EDTA and dextran)

----

fibrinogen treatment; but some losses can occur if fibrin formation takes place. Dextran has been widely used [30-34], and
Minor & Burnett [35] have reported that the increase in sedimentation is proportional to the dextran mol. wt. Since viscosity
is also a function of mol. wt., the range has to be carefully
chosen to achieve optimal sedimentation. Usually dextrans of
mol. wts. between 140,000 and 230,000 achieve the best separations, but often additional procedures are still necessary to
reduce the red-cell contamination to acceptable limits. Some
dextrans contain impurities which affect leucocyte metabolism,
but in most circumstances these can be removed by passage through
a charcoal column. With most of these agents for sedimentation
enhancement the best temperatures for sedimentation (8-15°) often
conflict with the optimum conditions (4-6°) for maintaining the
leucocytes metabolically competent. Moreover, the procedures
are often sufficiently lengthy to cause some concern about the
possible influence of proteases on the experimental findings.
Ficoll, used either alone or with some other agent such as
Hypaque, has been widely cited. It is fairly non-toxic to cells,
contains no ionizing groups and is of low osmotic activity.
Perhaps one of the best procedures is that described by Noble &
Cutts [36], in which the erythrocytes are first removed by a preliminary centrifugation over a single layer of 35% Ficoll. The
leucocytes can be harvested as a single band, and further resolution of this mixed population can then be achieved by linear
gradient centrifugation, likewise using Ficoll the strength of
which is varied to suit the experimental requirements (Fig. 1).

LEUCOCYTE ISOLATION FROM EXPERIMENTAL ANIMALS

PERITONEAL EXUDATE CELLS

ANIMALS	Rats, mice, rabbits and guinea-pigs
MATERIALS INJECTED	Glycogen, starch, high concentrations of glucose, casein, peptone, gum acacia, mineral oils, turpentine, liquid paraffin, aleuronal
INTERVAL BETWEEN INJECTION AND HARVESTING	4-20 h for polymorphs 4-7 days for macrophages

The commonest procedure for obtaining reasonably large quantities (0.5-1.0 g wet wt. of cells/animal) depends upon injecting irritant materials, usually intraperitoneally, to generate an inflammatory response. Diverse materials have been used for the initial insult to the animal (for examples, *see panel*). In general the sterile material is injected directly into the peritoneal cavity, and the cells are subsequently collected by peritoneal lavage using saline or isotonic phosphate solutions. The type of cell obtained depends not only on the species of animal and the nature of the agent injected but also on the elapse time before the exudate is removed. With most materials, with the exception of the mineral oils, the cells harvested between 4 and 18 h after injection are predominantly neutrophils. Populations enriched to the level of 90-95% polymorphs can be achieved with the most appropriate agent and with care in the timing of removal. After 4 days or more and particularly with the mineral oils, the peritoneal exudate cells become predominantly macrophage in character; but contamination with other cells (monocytes and lymphocytes) is generally higher than with neutrophil production methods. A further consideration too is that the bone marrow of the animal is being abnormally stimulated to release cells which migrate rapidly to the inflammatory site. In this context the cells do not reach normal maturity and perhaps should be regarded essentially as stress variants. However, with carefully controlled conditions it is possible to harvest cells in a reasonably homogeneous cell pool and of sufficient metabolic integrity to display energy-dependent motile behaviour and phagocytic activity, and also to show the full complement of enzyme activities associated with these phenomena.

PLATELETS

Amongst the blood formed elements, platelets are perhaps the easiest to separate free from contamination by other cell types. They are the smallest cells (2-3 μm in diameter) with densities ranging between 1.039 and 1.059 g/ml, lower than those for all other cells present. The choice of anticoagulant is most

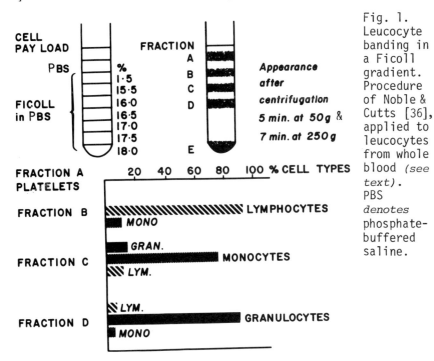

Fig. 1. Leucocyte banding in a Ficoll gradient. Procedure of Noble & Cutts [36], applied to leucocytes from whole blood (see text). PBS denotes phosphate-buffered saline.

important here in relation to the kind of investigation one intends to make. Even with anticoagulants present, silicone-treated glassware or plastic vessels are an essential require-emtn for the maintenance of shape and full metabolic and functional activities. In the circulation, platelets are lentiform in shape and when they are stimulated as in the haemostatic process in vivo, they change rapidly from discs to spheres. This shape change is followed rapidly by pseudopodia formation, and then irreversible changes occur coincident with the release or loss of intragranular stored materials. Some of the additives used to inhibit these changes taking place inadvertently (apyrase, caffeine, prostaglandin E_1) have been referred to earlier. The simplest procedure for isolating platelets from blood of man and most animal species is by low-speed centrifugation (130-170 g for 15-20 min), removal of the platelet-rich plasma, and 'high'-speed centrifugation for 15-20 min (>200 g) to pellet the platelets. By repeating once or twice more the cycle of low and 'high'-speed centrifugation on the re-suspended pellet, it is possible to harvest platelets with red-cell contamination <1 in 10,000 cells and virtually free of leucocytes. The recovery of platelets with respect to whole-blood platelet counts corrected for haematocrit is never complete, but generally exceeds 80%. Another procedure which avoids the potentially damaging effect of pelleting platelets by high-speed centrifugation is that of

gel filtration [37-39]. Here the platelets in platelet-rich
plasma are passed through a column of Sepharose 2B which has an
exclusion limit of mol. wt. >40 ×10^6 for protein molecules, by
elution with a balanced salt solution of strength 295 m-osmol.
The platelets seem well preserved morphologically and react well
with aggregating agents, and the procedure has the advantage of
ease of standardization of washing procedures.

Platelets show a higher degree of sub-population heterogeneity than other blood cells, particularly with respect to size,
density, protein synthesis and capacity to participate in haemostatic behaviour. They have a life-span in the circulation of
around 5-8 days, and it is believed that the *in vivo* aging is by
a process independent of random destruction and that heterogeneity reflects senescence. There is a considerable consensus of
opinion that platelets separated on the basis of density represent age-specific sub-populations, and certainly in some studies
a correlation between density and volume has been shown — the
larger platelets being the heaviest [40]. Moreover, the heavy
platelets are more active metabolically and are also more responsive to haemostatic stimuli [41]. The understanding of platelet heterogeneity has been further complicated by the recent
findings [42] that platelets lose specific surface-membrane
components during their sojourn in the circulation, presumably
by reversible encounters with the cells of the blood-vessel
wall. The age-related nature of this process has not yet been
fully explored, nor the effect upon buoyant density.

CONCLUDING COMMENT

This survey, though less exhaustive than a full review of the
field, may serve to illustrate some of the many difficulties
which research workers face in studying blood-cell populations.
To date almost every author has devised his own experimental
conditions and procedures for evaluating them, and to say that
many of these have evolved from empirical and highly subjective
observations would not be an overstatement. Some rationaliza-
tion of these difficulties backed by sound experimental proto-
cols would certainly be timely.

References

1. Luus, H. (1976) in *Cell Separation Methods* (Bloemendal, H.,
   ed.), North-Holland, Amsterdam, pp. 55-65.
2. Diem, K. & Lentner, C. eds. (1970) *Scientific Tables, 7th
   edn.*, Geigy, Basle.
3. Zucker, M.B. (1974) *Thromb. Diathes. Haemorrh. 33*, 63-64.
4. Zucker, M.B. (1977) *Fed. Proc. 36*, 47-49.
5. Eika, C. (1972) *Scand. J. Haematol. 9*, 248-257.
6. O'Brien, J.R., Shoobridge, S.M. & Finch, W.J. (1969) *J. Clin.
   Path. 22*, 28-31.

7. Besterman, E.M.M. & Gillett, M.P.T. (1973) *Atherosclerosis 17,* 503-513.
8. Gillett, M.P.T. & Besterman, E.M.M. (1973) *Lancet iv,* 1204-1205.
9. Kuper, S.W.A., Bignall, J.R. & Luckock, E.D. (1961) *Lancet i,* 852-853.
10. Tullis, J.L. (1958) *New Engl. J. Med. 258,* 569-578.
11. Legge, D.G. & Shortman, K. (1968) *Brit. J. Haematol. 14,* 323-335.
12. Cutts, J.H. (1970) *Cell Separation,* Academic Press, New York.
13. Paul, J. ed. (1965) *Cell & Tissue Culture,* Williams & Wilkins, Baltimore.
14. Schwock, G. & Passow, H. (1973) *Mol. Cell. Biochem. 2,* 197-218.
15. Dodge, J.T., Mitchell, C. & Hanahan, D.J. (1963) *Arch. Biochem. Biophys. 100,* 119-129.
16. Hanahan, D.J. (1973) *Biochim. Biophys. Acta 300,* 319-340.
17. Marks, P.A., Johnson, A.B. & Hirschberg, E. (1958) *Proc. Natl. Acad. Sci. U.S.A. 44,* 529-536.
18. Westerman, M.P., Pierce, L.E. & Jensen, W.N. (1963) *J. Lab. Clin. Med. 62,* 394-399.
19. Hjelm,M. (1968) *Folia Haematol. 89,* 392-399.
20. Bishop, C. & Van Gastel, C. (1969) *Haematologia 3,* 29-41.
21. Winterbourn, C.C. & Batt, R.D. (1970) *Biochim. Biophys. Acta 202,* 1-8.
22. Dain, A.R. & Hall, J.G. (1967) *Vox Sanguinis 13,* 281-284.
23. Minor, A.H. & Burnett, L. (1948) *Blood 3,* 799-802.
24. Minor, A.H. & Burnett, L. (1949) *Blood 4,* 667-669.
25. Skoog, W.A. & Beck, W.S. (1956) *Blood 11,* 436-454.
26. Buckley, E.S., Gibson, J.G. & Walter, C. (1950) *Proc., Centrifugal Separation of Formed Elements,* Harvard Univ. Press, Cambridge, Massachusetts, p. 33.
27. Gray, S.J. & Mitchell, E.B. (1942) *Proc. Soc. Exptl. Biol. Med. 51,* 403-404.
28. Winzler, R.J., Williams, A.D. & Best, W.R. (1957) *Cancer Res. 17,* 108-116. *See also* Winzler, R.J. & Wells, W. (1959) *Cancer Res. 19,* 377-387.
29. Laszlo, J., Stengle, J., Wright, K. & Burk, D. (1958) *Proc. Soc. Exptl. Biol. & Med. 97,* 127-131.
30. Miller, R.R. & Ackerman, G.A. (1959) *Amer. J. Clin. Path. 31,* 100-102.
31. Walford, R.L., Peterson, E.T. & Doyle, P. (1957) *Blood 12,* 953-971.
32. Freireich, E.J., Judson, G. & Levin, R.H. (1965) *Cancer Res. 25,* 1516-1520.
33. Nelken, D. (1963) *Vox Sang. 8,* 638-640.
34. Lapin, J., Horonick, A. & Lapin, R.H. (1958) *Blood 13,* 1001-1005.
35. Minor, A.H. & Burnett, L. (1952) *Fed. Proc. 11,* 422.

36. Noble, P.B. & Cutts, J.H. (1968) *J. Lab. Clin. Med. 72,*
    533-538.
37. Tangen, O., Berman, H.J. & Marfey, P. (1969) *Circulation*
    40, suppl. 3, 201.
38. Tangen, O. & Berman, H.J. (1971) *Thromb. Diath. Haemorrh.*
    25, 268-278.
39. Tangen, O. & Berman, H.J. (1972) in *Platelet Function in*
    Thrombosis (Mannucci, P.M. & Gorini, S., eds.), Plenum,
    New York, pp. 24-26.
40. Karpatkin, S. (1971) *Ser. Haematolog. iv,* 75-97.
41. Vainer, H. (1972) *as for* [39], pp. 191-217.
42. George, J., Lewis, P.C. & Morgan, R.K. (1978) *J. Lab. Clin.*
    Med. 91, 301-306.

#D-4
METHODS FOR ISOLATION OF LYMPHOCYTES

M. J. OWEN and M. J. CRUMPTON
National Institute for Medical Research,
Mill Hill, London, NW7, U.K.

Sources

Free cells: *cultured lymphoblastoid lines, leukaemic cells, peripheral blood, thoracic duct lymph.*
Solid tissue: *lymph node, spleen, thymus, tonsil.*

Dissociation *Tissue minced/macerated in Eagle's BHK medium.*

Separation *Erythrocytes, granulocytes, monocytes/macrophages and dead cells removed on Isopaque-Ficoll.*

Product, and separation of B- and T-cells

Purified cells >90% lymphocytes, of which >98% viable by eosin dye exclusion.
T-cells: thymus, T-Cell lines (e.g. MOLT 4), nylon wool or anti-(Ig)-Sepharose (macrobeads) depletion, rosetting.
B-cells: B-cell lines (e.g. BRI 8), 'nude' mice, nylon wool or anti-(Ig)-Sepharose elution, rosetting.

Lymphocytes have been divided into two primary populations, namely T and B, according to their tissues of origin (thymus and bursa or bone marrow respectively), functions and surface markers [1]. Thus, T-lymphocytes play major roles in cellular immunity (e.g. they mediate graft rejection and delayed-type hypersensitivity as well as providing 'help' and 'suppression' in a humoral immune response), whereas B-lymphocytes are primarily identified with antibody production. T cells also possess surface markers which are not shared by B cells. The two most common surface markers used to distinguish T and B cells are the Thy-1 differentiation antigen and immunoglobulin. Thus, mouse and rat T cells express Thy-1 but lack surface immunoglobulin, whereas all B cells possess surface immunoglobulin but not Thy-1 antigen. B and T lymphocytes can also be distinguished by their differential sensitivities towards activation by mitogens. The plant lectin phytohaemagglutinin selectively stimulates T cells, whereas lipopolysaccharides from Gram-negative bacteria stimulate B cells only [2].

Populations of T and B cells are not homogeneous. At least three distinct subpopulations of T cells are known that mediate help, suppression and killing activities. They are distinguished by their different Ly (lymphocyte) surface antigens and in part by differences in their sensitivity to cortisone [3]. Subpopulations of B cells are also known. Thus, in mouse spleen a proportion of immunoglobulin M-bearing cells also possess immunoglobulin D on their surface. Undoubtedly, the recent development of a technique for raising monoclonal antibodies against cell surface components will result in the definition of further subclasses of both B and T lymphocytes on the basis of their characteristic surface antigens [4].

SOURCE OF CELLS

The choice amongst the diverse sources listed above depends primarily upon the requirements. Thus, cultured or leukaemic cells are ideal for the purification of surface antigens since they comprise a functionally homogeneous population of identical genotype and can in principle be cultured in unlimited numbers. These cells are, however, clearly unsuitable for any study of quiescent (unstimulated) lymphocyte function. Also, the expense of cultured cells detracts from their use in large quantities except where reproducibility of cell-type is crucial. In contrast, leukaemic cells can be obtained cheaply in large numbers by plasmaphoresis.

Lymphocyte suspensions can be conveniently prepared from solid tissue by mincing or teasing the tissue in a medium (e.g. Eagle's BHK medium containing 20 mM HEPES) or by puncturing the tissue and gently perfusing with medium using a syringe. The suspension comprises, in addition to cell clumps, separate live lymphocytes, numerous dead cells, erythrocytes, granulocytes, monocytes and macrophages. Several procedures have been employed to eliminate contaminating and dead cells. For example, sedimentation through albumin gradients has been used to separate dead and red cells from viable cells which are generally less dense [5]. An alternative method for the removal of red cells is selective lysis using 0.83% ammonium chloride for 10 min at 0-4°; under these conditions, lymphocyte structure and function are apparently not impaired [5]. The most widely used and undoubtedly the best method to free lymphocytes from other cells is centrifugation on Isopaque-Ficoll gradients. The cell suspension is layered onto a high-density medium (several commercial preparations are available) which separates cells by their differential sedimentation rates determined by size. Red cells, which are induced to aggregate by Ficoll 400 in the medium, dead cells and granulocytes sediment to the bottom of the tube whereas live lymphocytes collect at the interface. This method is used almost universally for purification of lymphocytes from cell suspensions,

although a possible disadvantage is that it can result in a preferential loss of T cells [6]. Practical details of the technique are well documented in a company booklet [7]. Lymphocytes
separated on Isopaque-Ficoll gradients are >98% viable as judged
by staining with vital dyes such as eosin and Trypan blue.
Although dead cells stain more intensely with Trypan blue, this
dye underestimates the percentage of dead cells and eosin is preferable (authors' unpublished observations).

Preparation of lymphocytes from blood involves their separation from the vast excess of erythrocytes. This is achieved by
centrifuging blood diluted with 2 vol. of RPMI 1640 medium (8 ml)
on 2 ml of Isopaque-Ficoll for 30 min at 400 g and 20° [6].
Alternatively, red cells can be removed prior to Isopaque-Ficoll
centrifugation by mixing 3.5% 250 dextran (16 ml) with blood
(20 ml) and incubating at 37° for 15-30 min. Under these conditions the red cells aggregate and sediment, leaving the lymphocytes in the supernatant.

Choice of the source of lymphocytes depends primarily upon the
subsequent requirements. If pure populations of B or T cells are
required it may be necessary to fractionate further the cell suspension (see below). Cultured cell lines or naturally occurring
lymphoid tissue can, however, provide distinct lymphocyte sub-
populations without the need for further fractionation. For
instance, the majority (95-99%) of human chronic lymphocytic
leukaemias are of B-cell lineage as judged by their expression of
surface Ig, Ia antigens, C3 and Fc receptors as well as their
responses to mitogens. These leukaemias provide a convenient
source of large quantities of human B cells which can be readily
separated by plasmaphoresis. Although these cells cannot easily
be adapted to culture, human B lymphoid cell lines (e.g. BRI 8)
have been established by transforming blood lymphocytes with
Epstein-Barr virus. In contrast, human T cell lines (e.g. MOLT 4)
are derived from acute lymphoblastic leukaemias. These cell
lines grow to densities of up to 2×10^6/ml in culture. In the
mouse, most lymphoproliferative diseases involve T cells whereas
mouse leukaemias initiated by transformation with Abelson virus
possess B cell characteristics.

Murine T cells can be prepared from thymus and B cells from
congenitally athymic ('nude') mice [8]. Similarly, avian T cells
are obtained from thymus and B cells from the bursa of Fabricus
[9]. Experimental manipulation is also used to provide specific
lymphocyte subpopulations. Thus, splenic lymphocytes from mice
which have been neonatally thymectomized, X-irradiated and reconstituted with either bone marrow (lymphocyte stem cells during
adult life) or foetal liver (stem cells during embryonic development) provide essentially pure populations of B cells.

MAINTENANCE AND CULTURE

The continuing 'health' of separated lymphocytes depends to a large extent on the medium or buffer into which they are re-suspended. In general, the more complete the medium the greater the preservation of viability. Thus, viability decreases more rapidly in phosphate-buffered saline than in the presence of balanced salts, and in incomplete (minus foetal calf serum) compared with complete medium. Also, viability decreases more rapidly at 0° than at room temperature or 37°. Two media commonly used are Eagle's BHK medium and RPMI containing bicarbonate buffer or a non-toxic organic buffer such as HEPES. With medium containing bicarbonate buffer a high CO_2 pressure must be maintained in order to prevent the medium becoming alkaline. The more complete medium RPMI 1640 is used in preference to Eagle's BHK for long term culture of cell lines and for isolation of human lymphocytes which are particularly sensitive to buffer conditions.

For culturing, media are usually supplemented with serum (10-20% v/v) and the antibiotics penicillin (100 units/ml) and streptomycin (50 µg/ml). Foetal calf serum is commonly used in small-scale culture, but in mass culture pig and horse serum are often employed. Serum is, however, not essential. For example, stimulated lymphocytes have been cultured in the absence of serum in RPMI 1640 containing additional glutamine (4 mM) [10], and the serum has been replaced by albumin, transferrin and lipids in the culture of lipopolysaccharide-stimulated mouse B lymphocytes [11]. In the latter case the addition of thymus 'filter' cells has permitted growth with as few as one B lymphocyte per culture [11]. Culturing in serum-free medium has the marked advantage of simplifying studies of lymphocyte function by preventing interference from undefined serum components.

SEPARATION OF B AND T LYMPHOCYTES

Frequently it is necessary to separate B from T cells as well as functionally-specific subpopulations from lymphocyte suspensions. Various separation methods are available (Table 1), and the choice depends primarily upon the actual subpopulation required. Table 2 lists markers, which can in principle be made the basis for these separations.

Many separation techniques are based upon differences in physical parameters. Buoyant density separation on albumin gradients is not suitable as a general procedure for separating mouse and human B and T lymphocytes, since these populations overlap extensively. However, it has been employed successfully for isolating different subpopulations and stages of differentiation of B or T cells using either continuous or step albumin

Table 1. Sources of B and T lymphocytes.

Abbreviations used: TXFL, thymectomized, X-irradiated and foetal
liver reconstituted; TXBM, thymectomized, X-irradiated and bone
marrow reconstituted; 'nude', congenitally athymic (nu/nu); E,
unsensitized sheep erythrocytes; EA, erythrocytes sensitized with
anti-(red blood cell) sera; EAC, erythrocytes sensitized with
IgM antibody and complement; Ig, immunoglobulin.

The citations are of articles giving sufficient methodological
detail to enable each technique to be reproduced.

Method	Type of lymphocyte	Species	Yield	Ref.
A. Use of specific tissue				
1. Leukaemic cells	B and T	human, mouse	-	6
2. Cultured cells	B and T	human, mouse	-	6
3. Thymus	T	all	-	8
4. Spleen	B	TXFL/TXBM/ 'nude' mice	-	8
5. Bursa of Fabricus	B	birds	-	9
B. Use of defined surface markers				
1. Rosette sedimentation: E	T	human	} variable{	6
EA, EAC	B	all		16,17
2. Column fractionation, Sephadex-anti-(Ig)	B	all	high	18
3. Fluorescence-activated cell sorting	any subset for which a characteristic surface marker is available	all	high	20,21
C. General surface properties				
1. Nylon wool adherence columns	T	all	variable	14,15
2. Charge separation	B and T	rodents	high	13

gradients (for details, *see* [12]). Free-flow electrophoresis
has been employed for separating both quiescent and stimulated
B and T lymphocytes in mouse, rat and guinea pig [13] (cf. H.-G.
Heidrich, *this vol.*). Although the technique gives less overlap
between human B and T lymphocytes than is obtained with density
separation, the closeness in electrophoretic mobility precludes
fractionation on a preparative scale. The method has a moderate

capacity and can fractionate about 6×10^7 cells/h. The main disadvantage is the high cost and specialized nature of the equipment.

Adherence to nylon fibre columns has been used as a routine method for preparing human and animal enriched T cell populations depleted of B cells, monocytes and macrophages [14]. With optimal conditions and recycling, 95% of eluted cells possess T-cell characteristics and are functionally active. The major limitation of the method is the poor and somewhat variable yield of T cells (50-60%). Although the loss of T cells on nylon fibre columns is partly non-specific, some T-cell subsets may be differentially retained. Thus, in the mouse, IgG Fc receptor-positive, theta-positive T cells are selectively retained on nylon fibre columns. Cells enriched in B lymphocytes can be squeezed off nylon fibre columns [15]. However, the enrichment is poor (∿2-fold), making this method generally unsuitable for preparing purified B lymphocytes.

Table 2. Lymphocyte surface markers. Some surface markers are not present on all lymphocytes of a particular type. Thus B lymphocytes may not have IgD [see refs. 17 & 27, inter alia].

Marker	Lymphocyte distribution B	T
IgM	+	-
IgD	+	-
C3 receptor	+	-
Fc receptor	+	±*
Ia	+	±†
SRBC receptor	-	+
TL	-	+
Ly-1	-	+
Ly-2	-	+
Ly-3	-	+
Ly-4	+	-
Ly-5	-	+
Thy-1	-	+

*Fc receptors are known to be present on some T cells.
†Ia antigens are present on some mouse, but apparently not human, T cells.

Techniques dependent on surface specificities

A battery of separation techniques exploit surface-receptor and antigenic differences between lymphocyte subpopulations. Rosette sedimentation exemplifies this approach [16, 17]. Thus, T cells can be selectively removed by virtue of their capacity to bind heterologous red blood cells, the resultant erythrocyte rosettes being separated from the non-rosette forming cells (i.e. B cells) by their preferential sedimentation on Isopaque-Ficoll. B-lymphocytes can be isolated as EAC (erythrocyte-antibody-complement) rosette-forming cells using erythrocytes (usually ox) that have been pre-sensitized with IgM antibody and complement. Under these conditions, B cells form rosettes *via* their C3 receptors. Similarly, EA (erythrocyte-antibody) rosettes are formed by Fc receptor-bearing B cells and erythrocytes pre-sensitized with homologous antibody. The yield of cells by rosetting is variable but can be high. An advantage is that both B and T lymphocytes can be recovered from the same initial population. The main disadvantages are the method's low capacity and the possible detriment to lymphocyte function of the procedures used to remove the bound erythrocytes (e.g. lysis by NH_4Cl).

T and B lymphocytes have also been separated by affinity chromatography on columns of antibody against immunoglobulin attached to Sepharose [18]. In order to avoid adsorption of cells *via* their Fc receptors, the $(Fab')_2$ fragments of the antibody are commonly employed as the adsorbent. Immunoglobulin-bearing cells are retained by the column and can be recovered by elution with free immunoglobulin, whereas the unadsorbed fraction is composed predominantly of T cells with a small proportion of null cells (Ig-negative, E-rosette-negative subset). The eluted fraction generally consists of more than 98% B cells as assessed by surface markers and functional properties. Recovery of B cells has been claimed to be quantitative in the human system, but we have found elution of pig B cells difficult to achieve, presumably due to the high affinity of the antibody against immunoglobulin used. This immunoadsorption method suffers from two disadvantages: firstly, cells might be adsorbed by virtue of surface immunoglobulin bound to their Fc receptors, and secondly, B cells adsorbed to anti-Ig columns may undergo membrane perturbation which could affect subsequent functional assays. Degalan glass-bead, anti-Ig columns have also been used to separate B and T cells [19]. The purity and yield of the separated cell populations does not, however, compare favourably with the Sephadex anti-Ig method.

A recent technique which is proving of particular value for separating lymphocyte subpopulations is cell sorting using a Fluorescence Activated Cell Sorter (FACS). The instrument and technique are well described by Herzenberg [20, 21]. Cells are sorted according to the degree of fluorescence after labelling with a fluorescently-tagged antiserum. In principle, any subpopulation of cells for which a specific antiserum against a surface structure is available can be thus separated. Anti-(Ig) and anti-(Thy 1) sera are obvious examples for B and T cell separation. Thus, Hunt *et al.* [22] used FACS to separate Thy-1-positive and Thy-1-negative fractions from rat bone marrow cells. The especial power of the technique lies, however, in its ability to separate subpopulations of B and T cells. With pig mesenteric lymph-node lymphocytes, Perles *et al.* [23] achieved separation of two T cell populations based on the differential binding of the

lectin phytohaemagglutinin, that differed in their growth response to PHA. FACS has been used in combination with the myeloma
hybrid antibody technique of Kohler & Milstein [24] to identify
subpopulations of rat lymphoid cells [4]. The technique has also
been employed in the isolation of specific antigen-binding B
lymphocytes [25]. However, since the cell sorter is too slow for
direct processing of the large numbers of cells required for
isolating specific antigen-binding cells, an initial fractionation using hapten-gelatin was used.

Cell sorting is a sensitive technique capable of recognizing
as few as 3000 antigen molecules per cell. After processing,
the cells are >98% viable and the degree of cross-contamination
can be minimized by manipulating cell size as well as fluorescence parameters. Also, current instruments can separate cells
on the basis of two surface antigenic differences by using dual
fluorescence. The technique is unsuitable for processing the
large numbers of cells required for the isolation of surface
antigens, although cell numbers suitable for culturing or functional studies are attainable. Unfortunately, the expense of
the FACS rules out its use for all but the richest of institu-
tions.

One caveat, however, is that a cell population obtained using
a particular marker may be heterogeneous with respect to other
markers. For example, Scher *et al.* [26] have shown that murine
B lymphocytes possess varying densities of surface Ig and differ
in their expression of the complement (C3) receptor. Separation
of Fc receptor-, C3 receptor- and surface Ig-bearing lymphocytes
has also been observed in rat thoracic duct lymphocytes [17].
Thus, claims for homogeneity of lymphocyte subpopulations should
not be made on the basis of selection by one surface marker only.

References

1. Greaves, M.F., Owen, J.J.T. & Raff, M.C. (1973) *T & B Lymphocytes: Origins, Properties and Roles in Immune Responses,*
   American Elsevier, New York.
2. Wedner, H.J. & Parker, C.W. (1976) *Prog. Allergy 20,* 195-300.
3. Cantor, H. & Boyse, E.A. (1977) *Immunol. Rev. 33,* 105-124.
4. Williams, A.F., Galfre, G. & Milstein, C. (1977) *Cell 12,*
   663-673.
5. Shortman, K., Williams, N. & Adams, P. (1972) *J. Immunol.
   Meth. 1,* 273-287.
6. *Report of a WHO/IARC-Sponsored Workshop on Human B and T
   Cells* (1974) *Scand. J. Immunol. 3,* 521-532.
7. *Ficoll-Paque for* in vitro *isolation of Lymphocytes* (1975)
   Pharmacia Fine Chemicals, Uppsala.
8. Raff, M.C. & Wortis, H.H. (1970) *Immunology 18,* 931-942.

9. Ragland, W.L., Pace, J.L. & Doak, R.L. (1973) *Biochem. Biophys. Res. Commun. 50*, 118-126.
10. Hesketh, T.R., Smith, G.A., Houslay, M.D., Warren, G.B. & Metcalfe, J.C. (1977) *Nature (Lond.) 267*, 490-494.
11. Iscove, N.N. & Melchers, F. (1978) *J. Exp. Med. 147*, 923-933.
12. Shortman, K. (1976) in In vitro *Methods in Cell Mediated and Tumour Immunity*, Vol. 2 (Bloom, B.R. & David, J.R., eds.), Academic Press, New York, pp. 267-277.
13. Hayry, P., Nordling, S. & Andersson, L.C. (1976) *as for* 12., pp. 309-318.
14. Julius, M.H., Simpson, E. & Herzenberg, L.A. (1973) *Eur. J. Immunol. 3*, 645-649.
15. Handwerger, B.S. & Schwartz, R.H. (1974) *Transplantation 18*, 544-547.
16. Parish, C.R. & Hayward, J.A. (1974) *Proc. R. Soc. Lond. Ser. B. 187*, 47-63.
17. Parish, C.R. & Hayward, J.A. (1974) *Proc. R. Soc. Lond. Ser. B. 187*, 65-81.
18. Chess, L. & Schlossman, S.F. (1976) *as for* 12., pp. 255-261.
19. Wigzell, H., Sundqvist, K.G. & Yoshida, T.O. (1972) *Scand. J. Immunol. 1*, 75-87.
20. Loken, M.R. & Herzenberg, L.A. (1975) *Ann. N.Y. Acad. Sci. 254*, 163-171.
21. Herzenberg, L.A., Sweet, R.G. & Herzenberg, L.A. (1976) *Scientific American 234*, 108-117.
22. Hunt, S.V., Mason, D.W. & Williams, A.F. (1977) *Eur. J. Immunol. 7*, 817-823.
23. Perles, B., Flanagan, M.T., Auger, J. & Crumpton, M.J. (1977) *Eur. J. Immunol. 7*, 613-619.
24. Kohler, G. & Milstein, C. (1975) *Nature (Lond.) 256*, 495-497.
25. Nossal, G.J.V., Pike, B.L. & Battye, F.L. (1978) *Eur. J. Immunol. 8*, 151-157.
26. Scher, I., Ahmed, A. & Sharrow, S.O. (1977) *J. Immunol. 119*, 1938-1942.
27. Abney, E.R., Cooper, M.D., Kearney, J.F., Lawton, A.R. & Parkhouse, R.M.E. (1978) *J. Immunol. 120*, 2041-2049.

#NC Notes and Comments
related to the foregoing topics

Editor's explanation. — This section comprises supplementary contributions together with some discussion remarks made at the Forum which led to this volume.

FEATURES OF ISOLATED LIVER CELLS — Section #A

Comments on #A-1   H.A. Krebs *et al.* —HEPATOCYTE COMPETENCE

D.N. Skilleter *put a query.—* How useful, as a criterion for the degree of cell damage, is the stimulation of respiration by succinate in freshly isolated liver parenchymal cells ? *Reply by* H.A. Krebs.— Although succinate as a small molecule should readily enter cells if the membrane is damaged, in practice it appears to be excluded, like the much larger molecule of trypan blue. *Comment from* D.L. Knook: with this criterion there can be an enormous response if a cell preparation is contaminated with subcellular components, which light microscopy will not reveal.

H.A. Krebs, *answering* D.J. Morré.— The explanation remains elusive for the long-known phenomenon that the depletion of adenine nucleotides found with liver slices is not seen with regenerating liver or hepatocytes. *Remark by* C.N.A. Trotman.— Insofar as intracellular mitochondria can survive serious cellular damage such as toluene produces, one wonders whether measurements of oxygen consumption are indeed informative.

Comments on #A-2   W.H. Evans & M.H. Wisher - SURFACE PROPERTIES

W.H. Evans, *in reply to questions.—* (R. Coleman) Can canalicular-type structures be seen in isolated hepatocytes ?— Yes; our hepatocyte sections [Fig. 1b in text] do show surface regions with such morphology, whilst other workers find by scanning e.m. that isolated hepatocytes are covered predominantly by microvilli. (C.C. Widnell) Is the decreased s.a. of p.m. ecto enzymes with cells isolated in the absence of trypsin inhibitor due to enzyme inactivation, p.m. loss during processing, or less clean fractions ?— The benefit of including a trypsin inhibitor together with collagenase probably results from a number of factors, but a simple answer is precluded since knowledge is lacking about the molecular nature of the tissue dissociation process and inter-relationships between the p.m. and intracellular membranes of isolated hepatocytes. We found that exposure of p.m. to trypsin at low concentrations actually increased 5'-nucleotidase and phosphodiesterase activities, whereas they fell with collagenase. Addition of a trypsin inhibitor had no influence on protein recovery in the p.m. fractions or on fraction purity. By 'PAGE' on p.m. preparations, glycoproteins showed less intense staining if from hepatocytes than if from liver tissue [*our ref*. 5]. So inclusion of the inhibitor may help protect the hepatocyte surface.

R. Coleman, *answered by* W.H. Evans.— Could p.m. endovesicles (cf. Seglen, #B-1) account for apparently low recovery of 5'-nucleotidase ?— Yes; intracellular enzyme may be lower in hepatocytes (which are less active in secretion) and not recovered in p.m. fractions. Assay with detergent was not tried, nor (*answer to* R.H. Hinton) with pronase.

Comments on #A-3   T.J.C. van Berkel —CELL-TYPE ENZYMOLOGY

In reply to C.C. Widnell.— For NPC compared with PC, the s.a. of adenyl cyclase and 5'-nucleotidase is comparable per unit area of p.m. although higher on a cell-protein basis (**cf**. cell voL: 4-fold greater for NPC); but in a total liver p.m. preparation the NPC p.m. could perhaps be identified by the glucagon-insensitivity of its adenyl cyclase. *Reply to* R.H. Hinton.— It is not known whether Kupffer and sinusoid endothelial cells differ in nucleotidase level. *Reply to* H. Glaumann.— In principle, NPC and PC could indeed be compared on the basis of acid-hydrolase content per lysosome. For NPC *vs*. PC: vol. contribution of lysosomes is x 10; relative enrichment of lysosomal enzymes is x 0.7-12, and if < 10 the enzyme concn. per lysosome will be lower for NPC - e.g. enzymes that hydrolyze the terminal carbohydrate moiety of glycolipids, *ref*. [28]. But such calculations must be interpreted with care because lysosomes in NPC are more heterogeneous than in PC.

General remarks by T.J.C. van Berkel.— Information is needed about the relation between leakage of cytoplasmic enzymes (often taken to reflect intactness) and other cell-health criteria; the known discrepancy between trypan blue uptake and enzyme leakage could imply a directional localization of the p.m. *Remarks by* G. Krack-Dubois.— We find **Erythrocin** B a better vital stain than trypan blue because it is less difficult to discern exclusion by phase-l.m. when serum is present in the medium, and we have found its exclusion to be well correlated with dead cells and LDH leakage. Moreover *(view expressed in a Forum talk)*, glycogen synthesis by adult-rat hepatocytes in the presence of glucose (5 g/1 allows synthesis in culture media for up to 4 h) is a good cell-health criterion; its synthesis and content are decreased by toxic substances such as carbon tetrachloride, paracetamol or nickel chloride.

Nomenclature of visualization reagents (EDITOR'S COMMENT)

Since "the continued use and abuse of trivial names for dyestuffs ....has led to much confusion" (leaflet from Reeve Angel Scientific), it would be commendable to use trivial names merely as alternates to the standard name and number to be found in the *Colour Index* (Soc. of Dyers & Colourists, Bradford, & Am. Ass. of Textile Chemists & Colorists). No policy has been enforced in the present book, even in respect of whether a capital first letter is used.

SEPARATION APPROACHES – Section #B

Comments on #B-1 P.O. Seglen –CELL DISAGGREGATION & SEPARATION
and further comments relating to Section #A

Comments on cell disaggregation.– (D.N. Skilleter:) The choice
of collagenase preparation may depend on the intended use of the
cells: (a) if for immediate incubation or assay of particular
constituents, we find that Sigma Type II collagenase is good in
respect of cell yield etc.; (b) if for primary culture, the Worthington enzyme seems more satisfactory. (R.J. Hay, *replying to
a question from* R.T. Robson:) Aspiration at the digestion stage
is essential for obtaining a single-cell suspension adequate for
counting, but in the context of recovering viable cells for culture it is generally regarded as detrimental. We usually mix
the cell suspension gently, using a pipette with a 1-2 mm bore
at the tip, to the minimum extent consistent with generating a
uniform cell suspension.

Comments on hepatocyte heterogeneity, mainly by D.L. Knook.–
One would expect a density-dependent separation to reflect known
heterogeneity of cell type (perhaps with density overlaps) according to position in the liver lobe, and also to ploidy – known to
be age-dependent (H.-G. Hilderson).

Remarks concerning cell intactness.– (D.L. Knook:) It is important
that there be e.m. checking of *ultra*structural intactness: this may
be poor even for cells that exclude trypan blue (e.g. organelle-
free zone in the cytoplasm, swollen mitochondria). The difficulty
in demonstrating typan-blue exclusion with cultured cells can be
overcome by adding serum to coat the cells. (H. Glaumann:) Mitochondrial disruption as revealed by e.m. is the clearest sign
that a cell is dead or dying with no possibility of recovery.
(C.N.A. Trotman:) The measurement of leaked enzymes (e.g. LDH)
may be a doubtful criterion since proteolytic enzymes from the
cells or introduced during preparation may inactivate them.

P.O. Seglen *(tenor of his replies to* H. Glaumann).– Oxygenation of the medium used during preparation is unnecessary, since
hepatocytes can tolerate quite extended periods of anoxia *(refs.*
[43,47] *in* #B-1). Whilst hepatocytes can be purified merely by
repeated sedimentations at 1 *g*, low-speed centrifugation saves
much time. *Ref. added by Editor.*– U. Hopf *et al.* [(1974) *Clin.
Exp. Immunol. 16*, 117-124] describe hepatocyte isolation from
human biopsy specimens after mechanical dissociation in Eagle's
medium with shaking, then filtration.

Comments on # B-2 D.L. Knook – SINUSOIDAL CELLS BY ELUTRIATION
              # B-3 W.S. Bont – UNIT-GRAVITY SEPARATION

D.L. Knook, *replying to* C.A. Price.– The largest amount of liver
material that can be handled by elutriation is normally 500×10^6

cells; but repeated runs can be performed since the procedure takes only 20-30 min. C.A. Price, *to* W.S. Bont.— A quantitative model is needed for the assumed dependence on viscosity of the stabilization of gradients against streaming. Others who have worried about the capacity of gradients in density-gradient sedimentation found the critical factor to be the ratios of the diffusion coefficients of the sedimenting particles to that of the density-gradient material.

J.F. Tait/W.S. Bont.— The apparatus described in #B-3 compares well, in respects such as cost, with that of Miller & Phillips [cf. 6], use of which takes more time for reasons of resolution.

Comments on #B-4   H. Pertoft *et al.* —SEPARATIONS IN PERCOLL

Activities of enzymes (e.g. SDH) tend to be high after organelle exposure to Percoll, perhaps reflecting stabilization ('fixation?') of lysosomal and other membranes by Percoll or possibly residual traces of PVP. Biological activities are a better criteria of the intactness of particles than criteria such as e.m. appearance. Transport processes may survive Percoll as judged by entry of the small RuBPC subunit into pea chloroplasts (N. Chua); *Acetabularia* chloroplasts isolated in Ludox gradients in B. Green's laboratory incorporated amino acids into protein only after the medium had been washed out. Membrane stabilization by Percoll may be so strong as to cause difficulty in releasing enzymes from lysosomes or malaria parasites from erythrocytes. A silica medium (Ludox AM) rendered the unstable chloroplasts from *Euglena* very stable to lysis, but not spinach chloroplasts. The observation of an unusually high proportion of non-viable cells in the separation of 'PC' and 'NPC' suggests a need to check Percoll for possible cytotoxicity. (*Editor's précis of points put forward; discussants included* C. Gilhuus-Möe, H.-G. Heidrich, H. Pertoft, C.A. Price, and E.M. Reardon.)

Interferences.— To remedy gelation and precipitation of Percoll when solubilizing protein (0.5 M NaOH) for Lowry estimation, a lower pH may be used, and a procedure sought to solubilize the protein at that pH; it may be helpful to largely remove particles from the silica medium. With appropriate blanks, Lowry estimation of protein presents no problem. (*Editor's précis of remarks by* H.-G. Hilderson, C.A. Price & H. Pertoft.)

Editor's Note re ELECTROPHORETIC APPROACHES *(cf.* #B-6 H.-G. Heidrich)

In studies which seem not to have been pursued, perhaps because of detriment to organelle material, a continuous-flow isoelectric focusing step ('transient-state electrophoresis') was tried in a separation scheme, e.g. for lymphocytes [Just, W.W., Joaquin, O. L.-V. & Werner, G. (1975) *Anal. Biochem. 67,* 590-601]. A system has been described for following the migration, in various modes of electrophoresis, of cells such as erythrocytes [Catsimpoolas, N. & Griffith, A.L. (1969) *Anal. Biochem. 69,* 372-384]. S. Hjertén (in *Cell Separation Methods,* ed. H. Bloemendal) discusses free-zone electrophoresis of cells.

#NC–1

A Note on
COUNTER-CURRENT CHROMATOGRAPHY FOR SEPARATING CELLS AND ORGANELLES BY DISTRIBUTION IN TWO-PHASE POLYMER SYSTEMS

IAN A. SUTHERLAND[1] and YOICHIRO ITO[2]
[1]National Institute for Medical Research,
Mill Hill, London NW7, U.K.
and [2]Laboratory of Technical Development,
National Heart, Lung and Blood Institute,
Bethesda, MD, U.S.A.

The separation of cells and cell organelles on the basis of their partition in two immiscible liquid polymer phases is becoming an established technique [1], as the separation is based on surface properties rather than merely size and density. E. Eriksson & G. Johansson (this vol.) give a comprehensive account of how to prepare 2-phase polymer systems and how to manipulate and optimize partition. They also review separations involving surface charge, hydrophobic partition, affinity partition and cell-cell interaction. Sometimes a degree of purification can be achieved in a single test-tube partition step; but the requisite purity or resolution may call for more lengthy multi-stage procedures such as counter-current distribution (CCD) to be used.

We have developed a new way of handling two-phase polymer systems that is more analogous to liquid-liquid chromatography than CCD in speed and method of operation, whilst not having the disadvantages of a solid support. It can suitably be called counter-current chromatography or phase partition chromatography.

The principle is best introduced by examining the behaviour of two immiscible liquids in a closed, glass coil that is slowly rotated at unit gravity. Whereas an air bubble or heavy bead trapped in the coil unit will travel to what we define as the 'head' of the coil by classic Archimedian action, the two immiscible phases themselves will remain equally distributed throughout the coil system (Fig. 1a), bubbling through one another to maintain their equilibrium. If at any time the coil is stopped (Fig. 1b), then the liquids will be distributed evenly throughout the coil. Any excess of either phase will accumulate at the 'tail' end of the coil. If the coil in our model is opened at both ends, then pumping of one of the phases into one end will result in the same phase emerging from the other. One phase is therefore mobile while the other is effectively stationary even though mixing is taking place. A slowly rotating open-loop coil at unit gravity therefore has all the requirements for liquid-liquid chromatography, a sample injected with the mobile phase

Fig. 1. The coil system: (a, *top*), distribution of the 2 phases during motion; (b, *bottom*), distribution when stopped.

being separated on the basis of partition between the mobile phase and the stationary phase (*either* phase can be mobile). This scheme has been tested in practice by Y. Ito [2] with aqueous/organic solvent systems.

The extension of this principle to handling 2-phase polymer systems is complicated by one of the very properties that makes these systems so desirable for separating almost identical materials, *viz.* the fact that they are so similar in density, and the interfacial tension between the phases is so small. The problem is overcome by slowly rotating the coil in an enhanced gravitational field, so that the polymer phases behave much like the aqueous/organic phases do at unit gravity (Fig. 1).

Apparatus designed on this principle has already been built [3] using rotating seals to get the phases to and from the slowly rotating coil. In practice, rather than the larger diameter glass coiled tube of Fig. 1, tightly wound helices of PTFE tubing are used with tubing diameters of 0.5 mm, 1 mm or 1.5 mm depending on the scale of separation envisaged. Ito has shown that resolution is proportion to coil length and inversely proportion to diameter [4]. Test separations on different strains of *E. coli* [5] using gradient techniques, and of mammalian red blood cells from different species utilizing partition [6, 7] have been reported that give the equivalent of 50-100 transfers using CCD techniques in less than 2 h.

For soluble material, or particulate material smaller than, say 1 μm, if some loss of resolution due to inferior mixing of the phases can be tolerated, use may be made of a stationary helical coil mounted around the circumference of a disc. In this case it is the pumping action alone that causes any mixing between the phases. This process, called *toroidal coil chromatography*, has successfully separated different strains of *E. coli* [8] and is now being applied to the separation of plasma membranes and also to purifying restriction-enzyme digests of DNA.

1. Albertsson, P.-Å. (1971) *Partition of Cells, Particles and Macromolecules,* Wiley, New York.
2. Ito, Y. & Bowman, R.L. (1978) *J. Chromatog. 147,* 221-231.
3. Ito, Y., Carmeci, P. & Sutherland, I.A. (1979) *Anal. Biochem. 94, in press.*
4. Ito, Y., Hurst, R.E., Bowman, R.L. & Achter, E.K. (1974) *Separation & Purification Meths. 3,* 133-165.
5. Ito, Y. & Sutherland, I.A. (1979) *A New Technique for Separating Cells and Macromolecules, 1, Bacteria Cells,* Proc. 1st Canad. Chromatog. Conf., Dekker, New York.
6. Sutherland, I.A. & Ito, Y. (1979) *A New Technique for Separating Cells and Macromolecules, 2, Mammalian Cells,* Proc. 1st World Chromatog. Conf., Dekker, New York.
7. Sutherland, I.A. & Ito, Y. (1979) *to be published.*
8. Sutherland, I.A. & Ito, Y. (1978) *J. High Resol. Chromatog. & Chromatog. Comm. 3,* 171-172.

I.A. Sutherland *replied thus to a question put by* R.J. Hay.— The device could cope with relatively large cells (e.g. cultured human cells or leucocytes), provided that they survive the rotating seal system. However, we are currently developing a Rotating Coil Rotor without rotating seals that will separate large cells, on the basis of their distribution in a 2-phase polymer system; leucocytes survive well, but other testing awaits our overcoming a few more engineering problems.

POPULATIONS FROM TISSUES OTHER THAN LIVER –
Section #C

Comments on   #C-1   R.G. Price – PREPARATIONS FROM KIDNEY
#C-2   C.N.A. Trotman – G.-I. MUCOSAL CELLS

Points put to R.G. Price.– (*Answer to* T.J.C. van Berkel) Even without a good recovery compared with the homogenate as needed for assessment of distributions, enzyme activities are of value, particularly where comparison is to be made with other approaches such as micro-dissection, provided that the cell types are well characterized morphologically. (*Answering* J.F. Tait) Isolated glomerular cells are capable of producing renin under certain experimental conditions, but in connection with possible survival of control (e.g. by K^+) one must remember that the connection to the distal tubule which is involved in the production of renin is lost when glomeruli are isolated. (*Answering* R. Hay) Concerning maintenance culture of isolated epithelial components, our data for rats are still preliminary; but cells from the proximal convoluted tubule of the bovine nephron have been successfully cultured [Cade-Treyer, D. (1975) *Ann. Immunol. 126C,* 201-218].

Replies by C.N.A. Trotman.–(*To* G.B. Cline) Persistence of mucus (affecting ^{14}C-leucine uptake?) is not a serious problem with rat material, much being eliminated by dilution and washing. Cells adhere when centrifuged; so to ensure uniformity the aliquots should be taken before washing. Rate sedimentation cannot succeed if the cells are clumped; pre-washing is undesirable and unnecessary since the cells move from the preparation-medium environment soon after layering onto the gradient. (*To* R. Hay, G. Maurer) In the protein synthesis studies *in vitro*, incubation of gastric cells with pentagastrin, dexamethasone or oestradiol had no short-term effect on the amount or spectrum of proteins synthesized. We have not studied whether drug uptake diminishes when the cells start to produce abormal proteins.

Comments on   #C-4   G. Raydt – PANCREATIC ISLET ISOLATION
#C-6   J.P. Luzio – ADIPOCYTE ISOLATION

H. Pertoft, *to* G. Raydt (cf. #B-4).– Islets are denser in Metrizamide than in Percoll because it binds to cells and tissues. Islets isolated in Percoll (not done with Metrizamide) have been successfully transplanted, into ob/ob mice.

Replies by J.P. Luzio.– (*To* J.F. Tait) Demonstration of the action of adrenaline in stimulating lipolysis needs albumin in the medium, to bind the fatty acid produced since a rise rapidly inhibits lipolysis. It is not known what role in adipocyte function is played by the vacuoles that are seen close to the p.m. (*answer to* R.H. Hinton). (*Remarks by* G. Siebert; *cf. next p.*) The usual survival time of adipocytes with counter-flow suspension seems to be 5-6 h in metabolic experiments. Unlike fat-pad material, subcutaneous adipose tissue (pig) is very rich in Na^+.

#NC-2

A Note on

ISOLATED ADIPOCYTES IN RELATION TO SOURCE

G. SIEBERT, E. SCHNELL and G. STURM,
Department of Biological Chemistry,
University of Hohenheim, Stuttgart, W. Germany.

Interest in adipocytes stems mainly from interest in subcutaneous adipose tissue or other sites of triglyceride deposition. In innumerable cases, epididymal adipocytes are investigated with the tacit assumption that they are representative of subcutaneous adipocytes. The validity of this assumption is questionable on the grounds of the following evidence (activities based on adipocyte wet wt. and protein). —

a) Acetyl-CoA carboxylase, widely regarded as the pacemaker enzyme in fatty acid biosynthesis, demonstrates the following activity ratios for subcutaneous *vs.* epididymal adipose tissue:

NZO mice, obese	60% *sucrose*		5.9
	37% *sucrose*		2.8
	37% *sucrose, biotin-deficient*		6.1
rats, fed ad			
libitum –	64% *glucose*		1.9
restricted			
feeding –	64% *glucose*		3.5
fed ad			
libitum –	32% *sucrose*		5.1

It follows that regulatory responses of this enzyme to different diets vary with the source of adipose tissue investigated.

b) When ^{14}C-D-fructose and ^{14}C-sorbitol are compared as precursors of triglycerides, ratios of specific radioactivities in glycerol *vs.* fatty acids are as follows:

ADIPOSE TISSUE SOURCE:	*SUBCUTANEOUS*	*EPIDIDYMAL*
sorbitol	*12.8*	*7.0*
fructose	*82.2*	*57.2*

We conclude from these data that utmost care is required before the epididymal fat pad is used as a model of subcutaneous fat.

#NC-3

A Note on

ISOLATION OF ACTH-RESPONSIVE CELLS FROM RAT ADRENAL CORTEX BY DENSITY-GRADIENT CENTRIFUGATION

F. UNGAR[1], J. HSIAO[1], J. M. GREENE[2] and D. R. HEADON*
[1]Department of Biochemistry, University of Minnesota Medical School, Minneapolis, U.S.A.
and [2]Department of Biochemistry, University College, Cork, Eire

An important prerequisite to the analysis of the mechanisms of hormone action is the purification of specific endocrine cells from heterogenous populations following tissue dispersion, as now described for adrenal cortex.

Source of cells
Adult-rat adrenal cortex, cleaned of fat and quartered, in groups of 20 (medulla removed by scraping).

Dispersion
Collagenase 25 mg, DNase 2.5 mg in 5.0 ml KRBGA†; 37°, 45 min, atmosphere 95% O_2-5% CO_2.

Separation
Filtrate from passage through a 100 wire mesh screen centrifuged at 1000 g, 10 min; pellet washed twice with 5 ml KRBGA; 2.0 ml aliquots from suspension in 5 ml KRBGA put on 15 ml linear Metrizamide gradients (10% → 40% w/v) in KRBGA: 7000 rev/min, SW 27.1 rotor in Beckman L5-65 centrifuge at 15°, giving 3 visible bands, which were collected using a Pasteur pipette.

Product
A middle band which was uniquely responsive to ACTH stimulation (Table 1), in 80% recovery by this criterion.

Isolated zona glomerulosa cells of the adrenal cortex have been prepared by Tait and co-workers [1] through 1 *g* sedimentation in bovine serum albumin gradients, the recovery being ∿30% of the cells loaded. Such methods enable the various cell types present in adrenal cortical tissue to be separated, but require longer periods of preparation and prove more expensive than those employed in this study. The present results demonstrate that Metrizamide can be successfully used in the preparation of intact functional cells from rat adrenal cortex.

1. Tait, J.F., Tait, S.A.S., Gould, R.P. & Mee, M.S.R. (1974) *Proc. R. Soc. Lond. B.* 185, 375-407.

Present address: Department of Biochemistry, University College, Galway, Eire   [formerly at the Cork address].
†KRBGA *denotes* Krebs Ringer-bicarbonate-glucose-albumin.

Table 1. Corticosterone response of separated cells after ACTH stimulation. Results are expressed as ng per 10^6 cells.

	ACTH, μUnits		
	0	20	100
Material loaded	180	200	600
From gradient:			
Top layer	0	0	0
Middle layer	450	500	900
Bottom layer	0	0	0

#NC-4

A Note on

ISOLATION OF ACINI FROM LACTATING RAT MAMMARY GLAND

D. H. WILLIAMSON and A. M. ROBINSON
Metabolic Research Laboratory,
Nuffield Department of Clinical Medicine,
Radcliffe Infirmary, Oxford, U.K.

Until recently the standard *in vitro* preparation for study of lactating mammary-gland metabolism was the tissue slice. Slices have the disadvantage that they are tedious to prepare, consist of a mixed population of cells, lose adenine nucleotides on in-cubation, and are not suitable for time-course experiments. An alternative is the perfused intact gland [1]; but this requires considerable technical expertise and is not suitable for wide-ranging metabolic experiments. The recent description of an isolated acinar preparation [2] has provided a suitable system to study mammary-gland metabolism. A modification of this method [3] is described here.

Source   *Inguinal mammary glands from rats at peak lactation (12-16 days post-partum with 8-12 pups).*

Dispersion *Glands (5-8 g) excized, finely chopped with a razor blade and the mince suspended in warm (30-37°) Krebs-Henseleit saline [4] in a 50 ml measuring cylinder. Mince washed by decantation × 3 with warm saline, transferred to a 250 ml plastic flask and suspended in 30 ml saline containing collagenase (1 mg/ml; Boehringer, Grade II). Time taken 4-5 min. Suspension gassed with O_2/CO_2 (19 : 1) and shaken at 37° in a Dubnoff-type shaker.*

Separation *After 60-80 min the viscous suspension is filtered
through nylon gauze (mesh size 0.44 mm; Woolworth
tea strainer), and the filtrate centrifuged in a
plastic tube at 400 rev/min for 1-2 min. Residue on
gauze removed and re-extracted × 3 with a total of
50 ml warm saline and then centrifuged. Combined
pellet (1-2.5 g wet wt.) washed × 4 with 30 ml of
warm saline (fat droplets removed with tissue paper)
and finally suspended in saline to give about 50 mg
wet wt./ml. Acini maintained in suspension with a
magnetic stirrer.*

Product *The metabolic properties of the acinar preparation
have been found to accurately reflect alterations in
gland metabolism in various physiological states [3].
Rates of substrate utilization and lipid synthesis
are linear for at least 90 min at 37°, and ATP is
maintained at a concentration approaching that found
in vivo [5]. The acini are responsive to insulin [5].*

Alternative *Alternative methods for dispersion of the mammary
methods tissue give a preparation consisting of individual
cells [6, 7] but lower yields.*

1. Mendelson, C.R. & Scow, R.O. (1972) *Amer. J. Physiol. 223,*
1418-1423.
2. Katz, J., Wals, P.A. & Van de Velde, R.L. (1974) *J. Biol.
Chem. 249,* 7348-7357.
3. Robinson, A.M. & Williamson, D.H. (1977) *Biochem. J. 164,*
153-159.
4. Krebs, H.A. & Henseleit, K. (1932) *Hoppe-Seyl. Z. Physiol.
Chem. 210,* 33-66.
5. Robinson, A.M. & Williamson, D.H. (1977) *Biochem. J. 168,*
465-474.
6. Martin, R.J. & Baldwin, R.L. (1971) *Endocrinology 89,* 1263-
1269.
7. Greenbaum, A.L., Salam, A., Sochor, M. & Mclean, P. (1978)
Eur. J. Biochem. 87, 517-524.

D.H. Williamson, *in answer to* J.S. Major.— In studies now being
extended to a range of hormones, insulin has shown a stimulatory
effect on the isolated acini, but prolactin is relatively ineffective.

#NC-5
A Note on
SEPARATION OF CELL TYPES FROM THE CENTRAL NERVOUS SYSTEM

R. BALÁZS, J. COHEN and P. L. WOODHAMS
MRC Developmental Neurobiology Unit,
Institute of Neurology,
33 St. John's Mews,
London WC1, U.K.

To obtain isolated viable cells the procedure to dissociate nervous tissue is even more critical than it is in the case of other organs. This is related to the geometry of neuronal cells, most of which have many elaborate processes. Enzymic digestion is most frequently used since the structural preservation is better than after mechanical disruption. A method that preserves both ultrastructure and metabolic activity [1-4] is now outlined.

Source   *Cerebella from developing rats (1-14 days after birth) were used either with or without hydroxyurea pre-treatment [3, 4]. Hydroxyurea (a DNA synthesis inhibitor) eliminates replicating cells that sediment heterogeneously, and thus allows an improved resolution of differentiated cells.*

Dissociation *Brief incubation with trypsin in low concentration (0.025%), terminated by adding a trypsin inhibitor.*

Separation   *Sedimentation at 1 g; separation primarily according to size [3]. Fractions pooled if >50% of one size class.*

Products   *By electron microscopy and immunochemical as well as combined metabolic and autoradiographic criteria, fractions enriched in the following cell types are identifiable (in parentheses: proportion of the dominant cell type in the fraction):-*
large neurones, mainly Purkinje cells ($\sim$80%); astrocytes ($\sim$50%); late S and G_2 replicating cells ($\sim$40%); external granules and differentiating internal-granule cells ($\sim$70%).
Cell recovery 26-43% depending on age; viability >80% judged by plating efficiency in monolayer cultures, which for cells such as large neurones and astrocytes showed signs of differentiation.

Comments   *The fractionation, which is primarily according to size [3], owes its success in furnishing cells in good condition to the combination of factors employed in the crucial process of tissue dissociation: the above trypsinization, the use of isotonic conditions and physiological pH throughout, and the avoidance of high centrifugal forces. The products allow biochemical properties of particular cell types to be studied.*

1. Wilkin, G.P., Balázs, R., Wilson, J.E., Cohen, J. & Dutton, G.R. (1976) *Brain Res. 115*, 181-199.
2. Cohen, J. & Balázs, R. (1977) in *Membranous Elements and Movement of Molecules* [Vol. 6, *this series*] (Reid, E., ed.), Horwood, Chichester, pp. 261-262.
3. Cohen, J., Balázs, R., Hajos, F., Currie, D.N. & Dutton, G.R. (1978) *Brain Res. 148,* 313-331.
4. Cohen, P., Woodhams, P.L. & Balázs, R. (1979) *Brain Res. 161,* 503-514.

R. Balázs, *in reply to a question from* G. Siebert.— Concerning amino acid transport in the two cell types, both showed Na$^+$-dependence. The differences seem to relate to the conditions which lead to collapse of the Na$^+$ gradient. Neurones as excitable cells are endowed with a specific action-potential Na$^+$ ionsphere, but the role of Na$^+$,K$^+$-ATPase in maintaining the gradient would be expected to be the same in the two cell types, whilst possibly differing in terms of ouabain sensitivity. Alternatively, Na$^+$ permeability may differ. We are currently investigating these questions.

For brain-cell separation, SEE ALSO P.Giorgi, p.241 *in Vol. 4 (Reid, E., ed.)*

Editor's Note: ISOLATION OF GROWTH-PLATE CHONDROCYTES

Chondrocytes have been isolated from epiphyseal (metatarsal) cartilage from 5½-week old rabbits [1]. The matrix was dissolved away by use of papain and collagenase, either (a) in low concentration with a long incubation period, or (b) in high concentration with a short incubation period. Later [2], cells thus isolated were examined either fresh or after slow freezing (with 10% DMSO present) and rapid thawing, with respect to incorporation of Na$_2$^{35}SO$_4$ as assessed by autoradiography (and also by isolating and counting proteoglycan). With (a) but not with (b), there was fair resistance to freezing-and-thawing damage, although the proportion of labelled cells was 30-40% lower than that of eosin-excluding cells. The proteoglycan counts confirmed these results.

1. Lowe, A.C. & Smith, A.U. (1975) *Lab. Pract. 24,* 511.
2. Lowe, C. (1976) *Lab. Pract. 25,* 392-393.

#NC-6
A Note on
**THE AMERICAN TYPE CULTURE COLLECTION, A UNIQUE
RESOURCE FOR RESEARCH AND EDUCATION**

ROBERT J. HAY,
Cell Culture Department,
American Type Culture Collection,
Rockville, Maryland, U.S.A.

The American Type Culture Collection is a private, non-profit
international resource, with well-equipped laboratories, dedica-
ted to the collection, preservation, and distribution of authen-
tic cultures of living microorganisms and animal cells. One
unique feature is its range: Total culture holdings within all
departments are now over 20,000 and the distribution in 1978 exceeded 29,000. More than one-half million ampoules of freeze-
dried or frozen living materials are inventoried in mechanical
refrigerators at -60°C (-76°F), in large walk-in cold rooms at
+5°C (41°F), or in vacuum-insulated refrigerators automatically
supplied with liquid nitrogen at -196°C (-320°F).

Laboratory efforts include not only studies designed to improve technology for characterization and preservation of cells
and microorganisms but also research programs which exploit the
novel character of the institutional collections and facilities.
Collaborative studies with visiting scientists interested in
utilizing the collection are encouraged. All activities are
supported by funds from the National Institutes of Health,
National Science Foundation, Department of Agriculture, Food and
Drug Administration, National Heart, Lung and Blood Institute,
National Cancer Institute, private industry, foundations and in-
come derived from culture fees.

The Cell Culture Department which is one amongst the ten
within the ATCC currently maintains for distribution 400 cell
lines from 43 different metazoan species [1]. Holdings include
over 220 strains of human fibroblasts derived from normal donors
or from individuals with one of 34 genetic or other disorders.
Many of the latter strains were set up within the Department.
All certified cell lines (CCL) distributed are subjected to a
comprehensive and rigid series of quality control tests with
regard to growth, purity and identity [1-3]. Representative CCL
holdings are listed in Table 1.

The foetal bovine serum used in preparation of culture media
(antibiotic-free) is screened at the ATCC for the presence of
mycoplasma and, in the Virology Department, of BVD virus. In
addition, sera are selected which promote high plating

Table 1. Representative certified cell lines in the ATCC repository.

See the ATCC Catalog of Strains for additional information [1].

* Conclusive evidence is available indicating that various human cell lines in common use are contaminant derivatives of the HeLa line. Biochemical and cytogenetic analyses have been published to document these findings.

ATCC No.	Designation*	Species & tissue of origin*	Culture morphology	Specific characteristics or uses	G-6-PDH* isoenzyme	HeLa* markers	Chromosome No. 2n	Chromosome No. Cell line
CCL 1	NCTC clone 929	Mouse connective tissue	Fibro.	One of the first established; used extensively in virology, tumor biology and cell physiology.			40	65-66
CCL 2	HeLa	Human, epitheloid cervical adenocarcinoma	Epith.	First epithelial line from human tissue. Used extensively in virology, tumor biology and cell physiology.	A	+	46	82
CCL 9.1	NCTC clone 1469	Mouse liver	Epith.	Nutritional metabolic and enzymological studies			40	41-42
CCL 10	BHK-21 (C-13)	Syrian hamster kidney	Fibro.	Virology, and cell transformation studies			44	44
CCL 13	Chang liver*	Human liver*	Epith.	Virology, biochemistry and tumor biology	A	+	46	70
CCL 34	MDCK (NBL-2)	Canine kidney	Epith.	Virology			78	78
CCL 61	CHO-K1	Chinese hamster ovary	Epith.	Nutritional studies and cytogenetics			22	20
CCL 79	Y-1	Mouse adrenal tumor	Epith.	Produce steroid hormones			40	41
CCL 81	Vero	African green monkey kidney	Fibro.	Virology			60	58
CCL 86	Raji	Human Burkitt lymphoma	Lympho.	Grows in suspension. No demonstrable EB virus	B		46	46
CCL 96	3T6	Swiss albino mouse embryo	Fibro.	Collagen and hyaluronic acid secreting			40	78,81
CCL 127	IMR-32	Human neuroblastoma	Fibro. & Neuro.	Neurology			46	49
CCL 131	Neuro-2a	Mouse neuroblastoma	Neuro.	Used for study of microtubular protein synthesis	B		40	94-98
CCL 144	MH_1C_1	Rat hepatoma	Epith.	Secretes albumin, and complement, conjugates and metabolizes testosterone and bilirubin			40	48
CCL 163	Balb/3T3 Clone A31	Balb/c mouse embryo	Fibro.	Contact sensitive, tumor biology			40	68

efficiencies in test cultures of human diploid and primary animal cells. All culture media are sterilized through a double series of 0.22 µm millipore filters (Triton-free) prior to use.

To determine whether a cell line is free from bacterial and fungal contamination, 7 different microbiological culture media are used. Test samples are incubated both aerobically and anaerobically at 26° and 37° for at least 2 weeks. Tests for the detection of mycoplasmal contamination include both the direct broth-to-agar culturing procedure, and the indirect Hoechst fluorochrome-staining technique, with an indicator cell system incorporating negative and positive controls.

Cell lines recently accessioned are also being screened for protozoan and viral contaminants. Tests in the former category have been performed in the Protistology Department and consist of routine inoculation of test cell cultures into ATCC media 400,711 and 807 [4]. Results are assessed regularly for at least 14 days with incubation at both 25° and 35°. Microscopic examinations are included. Assays for viral contaminants require inoculation into embryonated eggs, co-cultivation with selected indicator cell lines and haemadsorption tests with concomitant scrutiny for cytopathogenic effects.

Selected viral susceptibility tests have been performed on most cell lines as appropriate: e.g. the Vero line (ATCC-CCL 81) has been challenged with 19 different arboviruses to provide baseline data for its potential use as a cell substrate. Sensitivity of cell lines to polioviruses provides an indication as to primate *versus* non-primate origin, the latter being resistant.

The species of origin of a cell line is confirmed by indirect immunofluorescence antibody testing and by isoenzymological assay for lactic dehydrogenase and glucose-6-phosphate dehydrogenase [3]. Karyotypes also clearly distinguish species, and serve to provide additional characterization. Standard karyological maps are routinely constructed, and in many cases detailed karyotypes employing Giemsa-banded preparations have been included [2]. For human cell lines, quinacrine fluorescence studies are performed to detect the sex of the species.

Where pertinent and feasible, the evaluation of cellular structure and function is performed. Standard transmission electron microscopy has been employed in cases where specific organelles can best be identified thereby. Histological or histochemical tests are applied to verify presence of specific cellular constituents if these criteria are of importance for cell identification. Evaluation of cell function is somewhat problematical since the techniques required are often complicated and variable. To this date selected tests for specific lipid syntheses, steroid hormone release, gonadotropin and melanin production have been applied at the ATCC.

Table 2. Lung cell lines and lines of potential use for studies on pulmonary physiology available through the ATCC.
See the ATCC Catalog of Strains [1] for additional information.

ATCC no.	Designation	Cell type and source (amplified in Catalog [1])	Donating scientist(s)
		Presumptive type 2 origin	
CCL 149	L2	Adult rat	Kaighn
CCL 150	AKD	Foetal cat	Kniazeff
CCL 185	A549	Human lung carcinoma	Lieber
CCL 196	LA4	Mouse lung adenoma	Stoner & Theiss
		Fibroblastic morphology	
CCL 16,39	Don, Dede	Hamster (Chinese)	Hsu
CCL 40	Bu (IMR-31)	Buffalo	Dwight
CCL 64	Mv 1 Lu (NBL-7)	Mink	Kniazeff
CCL 75 CCL 75.1	WI-38 WI-38.SV40	Human embryo, diploid and hyperdiploid, transformed	Hayflick Girardi
CCL 88	Tb 1 Lu (NLB 12)	Bat	Kniazeff
CCL 95 CCL 95.1	WI-26 WI-26.SV40	Human embryo, diploid and hyperdiploid, transformed	Hayflick Girardi
CCL 100	GeLu	Gerbil	Macy
CCL 111	GL1	Gekko	Clark
CCL 134,135, 151,154, 190,191 & 199	LL24 to LL96A WI-1003 HLFa	Human postnatal, various ages, diploid: normal lung and lung from patients with idiopathic fibrosis	Crystal Hayflick Stoner
CCL 153	HFL 1	Human embryo, diploid	Crystal
CCL 158	JH4	Guinea pig	Hay
CCL 160	DBS-FRhL-2	Rhesus monkey	Petricciani
CCL 161	DBS-FCL-1	African green monkey (male)	Petricciani
CCL 162	DBS-FCL-2	African green monkey (female)	Petricciani
CCL 168	FoLu	Fox, adult, grey	Baer
CCL 171	MRC5	Human embryo, diploid	Jacobs
CCL 186	IMR90	Human embryo, diploid	Nichols
CCL 192	RFL-6	Rat	Macy
CCL 193	R9a	Rabbit	Early
CCL 194	DPSO 114/74	Squirrel monkey	Rangan
CCL 195	AHL	Hamster (Armenian)	Lavappa
CCL 197	NOR	Mouse*	Starcher
CCL 198	BLO	Mouse*	Starcher

*CCL 198 was isolated from muscle of mice with hereditary emphysema; CCL 197 = control.

Specific quality-control tests, to an extent depending on cell type, are performed on each cell line when it is first deposited, and again, more extensively, when the reference seed stock is prepared and whenever a seed stock ampoule is expanded to prepare a distribution freeze [5].

A Lung Cell Resource has recently been developed as part of the collection of cell lines with support from the NHLBI (N01-HR-6-2930). Its advisory committee includes Dr. Stephen Blose, Univ. of Pennsylvania; Dr. Ronal Crystal, NHLBI; Dr. Philip M. Farrell, Univ. of Wisconsin, Dr. M.E. Kaighn, Pasadena Foundation for Medical Research; Dr. Charles Kuhn, Washington Univ.; Dr. Elliot Levine, Wistar Institute; Dr. Robert Mason, Univ. of California; Dr. Donald Massaro, Univ. of Miami; Dr. Robert M. Senior, Jewish Hospital of St. Louis; Dr. Gary Stoner, NCI and Dr. Mary·C. Williams, Univ. of California.

Table 2 lists lung-cell lines currently available from the ATCC (cf. R.J. Hay, *this vol.*).

1. Hay, R.J., Lavappa, K.S., Macy, M.L., Shannon, J.E. & Williams, C.D. (1979) in *American Type Culture Collection Catalog of Strains II,* 2nd edn.
2. Lavappa, K.S. (1978) *Tissue Culture Association Manual 4,* 761-764.
3. Macy, M.L. (1978) *Tissue Culture Association Manual 4,* 833-836.
4. *American Type Culture Collection Catalog of Strains I,* 13th edn. (1978).
5. Hay, R.J. (1978) *Tissue Culture Association Manual 4,* 787-790.

POPULATIONS FROM NON-TISSUE SOURCES – Section #D

Comments on #D-1   E.M. Reardon *et al.* – MARINE MICROALGAE

E.M. Reardon, *replying to questions.*– (C. Schnarrenberger.) It is
not necessary to pre-concentrate dilute species by a risky procedure such as filtration or pelleting; with zonal flow-through
centrifugation, 200 1 of sea-water has been fed directly into a
gradient, and the algae collected in a 50 ml volume of gradient.
(R. Douce.) Whilst collected algal cells in principle could be
largely freed of Percoll by differential centrifugation, we had
earlier found that some of the more delicate species were damaged
by pelleting, and are glad to know of H. Pertoft's evidence that
Percoll does not interfere with a number of enzyme assays. (K.
Mueller.) As evidence of the purity of the separated populations,
the two algal species each gave a pure culture.

  Replies to G.B. Cline, R. Miller.– Algae survive at least 1 h
in Percoll; to ensure isopycnic sedimentation the organisms have
to be centrifuged for 20 min and the process then repeated. The
intention to use the CF-6 rotor hinges on the availability of
Percoll in large quantities as needed for continuous-flow centrifugations.

Comments on #D-2   S.M. Lanham – PARASITIC PROTOZOA

Replies to J.O. Molloy, C.A. Price.– We have not investigated
whether the viscosity of sucrose is detrimental to such highly
motile organisms as trypanosomes; but in the case of *T. cruzi*
preparations the organism is recovered from the top of the gradient at very low concentrations of sucrose. The short survival
time of merozoites, under 30 min, precludes separation by gradient centrifugation. Concerning efforts to bring *Plasmodium*
through a life cycle outside the hosts, much progress has been
made: the organism has been cultured up to the sexual stages.

VARIOUS SEPARATIONS: *Sundry refs. contributed by the Editor*

Cooper, P.H. & Stanworth, D.R. (1976) *Biochem. J. 156,* 691-707.
  - *Modified version of their method for rat (peritoneal) mast cells.*
Burghouts, J. *et al.* (1977) in *Cell Separation Methods* (Bloemendal,
  H., ed.), Elsevier/N. Holland, Amsterdam.- *Human immature myeloid
  cells (bone marrow) and G*$_1$ *& S-phase cells (leukaemic patients).*
Schulman, M.R. (1967) *Biochim. Biophys. Acta 148,* 251-255.-
  Rabbit reticulocytes fractionated in dextran density gradients.
Lindahl, P.E. (1958) *Nat. (Lond.) 181,* 784.- *Bovine spermatazoa
  carrying X- & Y-chromosomes, by counter-streaming centrifugation.*
Levine, S. (1956) *Science (Wash.) 123,* 185-186.- *Magnetic techniques
  for leucocyte isolation. Cf. the heroic procedure, now obsolete,
  for liver RES cells:* Wattiaux, R. *et al.* (1956) *Biochem. J. 63,* 608-612.
Cutts, J.H. (1970) *Cell Separation: Methods in Hematology,* Acad. Press.

FURTHER EDITORIAL CONTRIBUTIONS

Berg, T. & Munthe-Kaas, A.C. (1978) *Exp. Cell Res. 109,* 119-125.
- *Lysosomal enzymes in cultured rat Kupffer cells.*
Pretlow, T.C., *et al.* (1977) *Cancer Treatment Repts. 61,* 157-160.
- *Epithelial cell separation from prostate, normal & cancerous.*

*ALSO, in previous vols. in this series (Reid, E., ed; details oppos-
title p.): Vol. 3, Pasternak, C.A. & Warmsley, A.M.H.- Cultured
neoplastic mast cells in relation to cell cycle* (p. 249); Cline, G. B.
& Doggett, L.F.- *Fish eggs (and larvae) from plankton* (p. 253).
Vol. 4, Taylor, D.G. & Crawford, N. - *Centrifugation of platelets* (p.
319); Poole, R.K. & Lloyd, D., *Yeast-cell isolation at different
stages in the cell cycle* (p. 347).

ORGANELLE MARKERS: AN IMPORTANT COLLECTIVE VIEW

Although not strictly within the Cell Populations *theme of this
book, an outcome of vigorous debates at the Forum in July, 1978,
warrants mention, viz. a multi-author article* (Morré, D.J., *et al.)
which is expected to appear in* Eur. J. Cell Biol. — *from which
the following excerpts are reproduced with acknowledgement.—*

It would be useful, especially to editors and referees, to
have a 'code of practice'. Meanwhile, *Instructions to Authors* in
journals might include guidance on the following lines -
1) In subcellular studies, sufficient characterization of fractions should be presented to ensure that the results and interpre-
tations are soundly based. Even where it is merely a matter of a
single pelleted fraction prepared as the test material, appropriate marker assays should be performed with special reference to
testing for contaminating elements.

2) Recovery and specific activity data should be presented,
accompanied in analytical studies (e.g. for assignment purposes)
by balance-sheet information related to the starting material.
Morphological characterization is also recommended. Journals
which do not print electron micrographs might request supporting
morphological data to inform the referees.
3) Only where assurance of purity is presented is it permissi-
ble to use terms such as *nuclei* or *lysosomes,* as distinct from
crude nuclear-fraction or *lysosome-enriched fraction.*

(Assay procedures should be meticulously chosen and described.
They should be specific to avoid complications from two or more enzymes in different subcellular locations........All assays should be
conducted after elimination of latency......)

ooooooooooooooooooooooooooo

General Index

CORRECTIONS to Volume 7
∴∴∴∴∴∴∴∴∴∴∴

Table of Contents: to #D-9 authorship add:
KARIN KIESEL and KARL-ERHARD MÜLLER

p. 7, *Title: misspelling of* SENSITIVITY

p. 170: *delete last 3 lines (are on next p.)*

p. 226, COMMENTS: *alter p. 319 to p. 335*

pp. 264 & 265 *to be transposed.*

p. 293, *Title: misspelling of* CHROMATOGRAPHIC

Compound-type Index: shift Dopamine *from* #IIa *to* #IIb'.

NOTES

NOTES